CASTLEGUARD

CASTLEGUARD

DALTON MUIR · DEREK FORD

281964205

National Parks
Centennial

1885

1985

Centenaire des
parcs nationaux

Published by authority of
the Minister of the Environment
©Minister of Supply and Services Canada 1985
Available in Canada through
Authorized Bookstore Agents
and other bookstores
or by mail from
Canadian Government Publishing Centre
Supply and Services Canada
Ottawa, Canada K1A 0S9

Catalogue No. R62-223/1985E   Canada: $34.95
ISBN 0-660-11790-8   Other Countries: 41.95 (Cdn.)

Price subject to change without notice

Contents

Foreword

Simply because we have chosen distinctive, interesting areas as representative of our Canadian landscapes and ecosystems, many of the characteristics of our national parks are conspicuously displayed, even to the casual observer. The most popular of the parks flaunt their glorious scenery shamelessly before millions of annual visitors. Supplementing these spectacular landscapes is a broad range of tourist facilities: paved highways and hiking trails, campgrounds and lodges, interpretive programmes and eco-tour pamphlets.

At the other end of the spectrum are wilderness areas, accessible only to the hardy, adventuresome and relatively affluent: pieces of Canada characterized by solitude and unspoiled nature. Between these extremes are parks with special personalities – rivers for canoeists and archipelagos for small craft, sand dunes and beaches and tidal mud flats, rolling grasslands, winter trails, hot springs, waterfalls and avalanche zones, migratory-bird refuges.

There are thirty-one national parks and a thousand different ways to celebrate the Centennial of the National Parks of Canada. Though each park has been selected to represent a particular landscape or ecosystem, they are not homogeneous areas; each is a mosaic of elements complementing one another.

It is a particularly happy circumstance that prompted the commissioning of this book as a contribution to the Centennial. In readable but scientifically accurate prose written to complement the remarkable photography, the authors have described the evolution and character of an extraordinary complex of high-altitude features in Banff National Park, the Castleguard caves and valley, small in extent, scientifically unique.

In a very real sense, the caves are more remote and much less easily explored than our most northerly Arctic islands. Their subterranean environment is unlike any that we experience on the surface. The Castleguard Caves provide geomorphologists with a natural laboratory, in which to study the creation of calcite decoration, the chronology of climatic change and the effects of hydrology and hydrochemistry on cave formation.

This is neither a dry research report nor a straightforward account of cave exploration. The authors have brought to life the geological history of the valley and its array of underground mysteries, imbued the complex with a geographic personality and given us a lively biography of this fascinating part of our first national park. It should serve as a model of popularized science for describing distinctive elements of other worthy areas in our parks system.

I am honoured to be invited to write a foreword to this innovative and successful approach to describing landscape evolution in a high-altitude, mid-latitude environment. In 1980, The Royal Canadian Geographical Society co-sponsored and contributed to a major Castleguard caves expedition, led by Derek Ford. This volume gives us reason to be proud of our early association with the investigation of this remarkable labyrinth.

J. Keith Fraser
Executive Secretary
The Royal Canadian Geographical Society
Publisher, Canadian Geographic

Castleguard

We are in a high, glacier-rimmed mountain valley. In the valley is an alpine meadow rich with vegetation, and beneath us lies an enormous cave. The valley and the cave are part of the Castleguard complex in Banff National Park. They take their name from Castleguard Mountain, which overlooks the valley. The Castleguard complex – mountain, valley and cave – stands at the edge of the Columbia Icefield, the largest surviving glacier in the Rocky Mountains.

The valley is small, about 7 km/4 miles long, and not especially deep or wide. But it is unusual because it is so high above sea level at its mid-point, 2230 m/7300 feet, and it is cut off at both ends and left "hanging." The Saskatchewan Glacier sliced off the head of the valley and the Castleguard River obliterated the valley's lower end when it carved a much deeper valley of its own. At the glacier end of the valley there is very little vegetation; but as we travel down the valley we see more and more growth. By the time we reach the river we are in a forest.

In the middle of the valley is an open, lush tundra; this large patch of rich growth gives the area its unofficial name, the Castleguard Meadows. In midsummer the meadows are ablaze with brightly coloured flowers. Most of the flowers that flourish here grow only in high places, near glaciers or in the Arctic.

The valley is a microcosm, a vivid picture of the way landscapes have developed since the great glaciers scraped across the earth. At one end is barren rock, only recently uncovered by retreating ice; in the middle is a flowering tundra; at the other end, close to the river, is thick, dense forest.

Beneath the lush meadow lies the Castleguard Cave. The name is a bit misleading: this is no small hole in the earth, but an enormous complex of passageways in solid rock. Its sources are under the Columbia Icefield, where some of the tunnels are blocked by glacier ice and only a few metres of roof rock separate the cave from 300 m/1000 feet of solid ice. The cave passes beneath Castleguard Mountain and reaches the surface 400 m/1200 feet lower, among the trees at the lower end of the Meadows.

Long and straight, angled or twisted, steep or gentle, 18 km/10 miles of tunnels and passageways have been explored.

Beneath these passageways lie yet other caves. These lower caves have not yet been entered and may never be explored, for they are a huge "plumbing" system for the whole Castleguard complex. Water enters this second system through sinkholes and leaves it by way of the Big Springs, below the entrance to the charted cave. The water comes from rain, snow and glacial melt. The lower system carries away the rain and snow, but hazardous midsummer melt causes violent floods at the system's downstream end, which back up into the explored cave, flooding it. Scientists have measured the second system's water-handling capacity, to estimate its size, and they believe that the second system is larger than the explored cave. Moreover, the second system is enlarging by a volume equal to that of a typical suburban house each year.

The Columbia Icefield with its glaciers, the valley, the two cave systems and the mountain that gives the system its name together form a scientifically unique complex. Castleguard is a beautiful landscape, and a vivid example of the way this country developed during and after the ice age. It is a dynamic place, which incorporates the comings and goings of glaciers and of living things. Time has special meaning here.

Geologists think of earth's history in terms of very long spans of time. Today we are in the period known as the Quaternary ice ages, which have lasted about 2.5 million years. Repeatedly, during these ages, the land was buried under ice several kilometres thick and then was exposed again when the ice melted. The ice depressed the earth's crust, tore down mountains, filled the valleys and overwhelmed the plains. Nearly all of what is now Canada was smothered in flowing ice, and there was no life; there was only wind, sifting snow, ice and silence. At Castleguard the great ice age established the major surface features we see today.

Then, about fifteen thousand years ago, the climate began a cyclical warming and the ice began to melt. The water flowed away on the earth's surface, cut deep valleys, entered underground streams, which flowed through deep cracks in the rock, carved them more deeply and created caves. Canada's landscape was stark, barren, brown and sterile, newly shaped, like much of present-day Castleguard.

But even as the ice was retreating, living things began to arrive. Plants added colour; animals explored the new country. Before the melting ice had disappeared, the first humans migrated from continents to the west. They probably saw the beginnings of forests and prairies, and hunted the woolly mammoth, one of the great beasts of the ice age: their stone spear-points are now found among bones of the animals they killed for food. We know something of the cultures these people brought with them and how they adapted to their new surroundings. Two different lifestyles arrived with the people who explored the newly emerged landscape. Inuit, who emigrated from an arctic climate, stayed north of the forests. Ancestors of North American Indians longed for warmer landscapes; they moved southwards till they found forests and bountiful prairies.

Since the last major ice age, or glaciation, a Little Ice Age, or "neo-glacial" period, has occurred at Castleguard and is only now drawing to a close. For the past seven hundred years the hemisphere has experienced a slight cooling. In high, mountainous areas glaciers again advanced; only in the past few decades did they withdraw. The effects of the Little Ice Age are especially striking at Castleguard, with its grinding glaciers, struggling plant and animal communities and the unsuspected cave-world below the surface. At Castleguard we are close to our past, close to our place in the natural order. As we walk through the valley, we can see the origins of the country in which we live.

We pass through landscapes that unfold like a film of the past ten thousand years of history. Along the edges of the valley we find places where, even today, new landscapes are being uncovered. As the glacial ice retreats, vegetation exploits the new habitat. The Little Ice Age happened entirely within historic time, and at Castleguard it has been ending within the one hundred years since Banff National Park was created in 1885.

Few people experience first-hand the beauty of Castleguard, where two environments exist one on top of the other. During warm months the Castleguard Caves flood unpredictably, so they can be explored only in the winter, and only with the help of a well-organized expedition. The Castleguard Meadows are reached by a difficult climb up a glacier. Visitors are few. In this book, we wish to share our experience with all the people who will never travel to Castleguard. Perhaps your lives will be enriched by this photographic visit, our celebration of Castleguard.

The Two Landscapes of Castleguard

The surface landscape of mountains, glaciers, an icefield, waterfalls and a richly vegetated valley floor is like a review of the end of the last great ice age, and of the withdrawal of a more recent "Little Ice Age," which is still underway.

Below it all, an extensive cave landscape riddles the mountains and meets the bottom of the icefield. It was carved out of solid rock by the most yielding of natural substances, water.

1/BESIDE AN ICEFIELD AND A GLACIER

Only wind and falling water intrude upon stillness; only clouds and their shadows move. The clean, arctic-blue light illuminates an ice-age scene that is devoid of life. Among these spectacular landscapes lies a place like no other. We shall call it simply Castleguard, from a mountain, a river and some glaciers that bear the name. High meadows and a deep cave nearby add subtlety and mystery.

The Columbia Icefield, still covered by last winter's snow, lies across the land, its edge a white band. Immediately below is a darker ice mass, the upper reaches of the Saskatchewan Glacier. Nearly free of winter snow, its cracked and dirty surface melts as the glacier descends to warmer temperatures. Projecting from the left is the dark flank of Castleguard Mountain, still partly snow-covered.

In the foreground is a high twisting ridge of moraine: gravel and rocks that were deposited by the glacier when it was thicker and extended farther down the valley. The peak in the background is Mount Columbia; it was probably covered during the greatest ice ages.

Nearby, on land overlooking the Saskatchewan Glacier, a few living things struggle to survive.

2/ THE LOWER END
OF THE MEADOWS

From the sterile glacial landscape, a valley slopes gently to the south. It and the surrounding scenery are the heart of Castleguard. The valley floor, "the Meadows," supports increasingly complex vegetation towards lower levels. So many natural features compete for attention that the senses become saturated, while much more still lies ahead.

The forested south end of the Meadows drops abruptly into a deeper valley. The Meadows are therefore a "hanging valley," because both ends are sheared off. Below the lower meadow is a large river valley, and beyond the ridges of Mount Bryce and Watchman Mountain dominate the view. Ice masses clinging to the sides of the ridges break free during summer months and crash down the mountainsides like thunder.

Major weather systems often follow this valley, for it is part of the Thompson Pass through the main ranges of the Rocky Mountains. Rain and lightning storms often break over Watchman and Mount Bryce while the Meadows remain clear and dry.

3/A WET ROCK

In the forest below the Meadows is a grey rock; a shallow depression is still partly filled with water from last night's rain. At each end tiny cracks in the rock are being widened and deepened by the water. This dissolving process is the key to a vast underground landscape. For the water that eases our thirst and trickles softly through our fingers helped tear apart the peaks and carved a vast cave, which underlies these mountains.

A century ago, these landscapes were set apart to reward the searching eye, the exploring footstep and the receptive mind. Today their secrets are beginning to unfold.

4/ENTRANCE TO AN UNDERGROUND LANDSCAPE

Near the wet rock lies a cave entrance, the only way into a vast and complicated landscape carved out of limestone by water sinking from above.

Late each day during hot summer weather, high snow, the icefield and nearby glaciers melt so fast that the natural cave drainage is overwhelmed. Water backfloods, out of the caves below, and pours down the limestone pavement in the foreground. It sometimes washes completely over the huge boulder on the right.

This flooding seals the cave, sometimes for many days. During summer, only the first cavern may be entered safely. The other caverns and passages must be explored during winter months, when there is no danger of flooding.

Sitting in the cool breeze that comes out of the entrance chamber on a hot day, one finds it difficult to imagine what lies beyond. More difficult still is comprehending that the subtle processes visibly changing the surface of the nearby wet rock have carved the miles of cave that lie inside. Perhaps most difficult of all is the long wait for winter, when we may enter.

5/IN THE HEART OF CASTLEGUARD CAVE

Castleguard Cave is awesome and tantalizing. Because of the danger and formidable obstacles, it can be visited only by experienced cave explorers. This is the first glimpse of a landscape that even hikers to Castleguard can never hope to experience first-hand.

Only a few cavers have ever pushed through to the end, but their photographs, taken during research expeditions in the past eighteen years, allow us to share something of their time underground, where days and nights are indistinguishable, but cold, wet fatigue is ever-present.

This scene, beneath Castleguard Mountain in the middle of the cave, expresses some of the mystery of the place. Pristine galleries that have never known light or a human footprint beckon us on.

Beneath the known cave lie unexplored cave systems, which carry off most of the meltwater each summer. That these systems are at least as large as the known cave can be determined from the huge volume of water they carry, but these lower caves may never be entered.

Populating a New Land

The glaciers of the great ice ages covered much of the northern hemisphere. When the climate changed and the glaciers retreated, a sterile landscape was revealed. Vestiges of the great ice ages remain here among the high mountains, and living things still struggle to populate this harsh new environment.

6/CASTLEGUARD MOUNTAIN AND THE COLUMBIA ICEFIELD

Castleguard Mountain dominates the southeast sector of the Columbia Icefield. The icefield is huge, about 325 km²/127 square miles in area and about 300 m/1000 feet thick. It moves, slowly, towards the edges, where valley glaciers carry away the ice. The motion causes deep crevasses and meltwater channels to form. Some of these are seen as dark areas of dirty ice in the white snows of the previous winter. Cloud shadows mottle the icefield, while Castleguard Mountain is in sunshine.

On the icefield there is no life. Ice-age conditions have continued here for at least 170,000 years.

7/THE COLUMBIA ICEFIELD

Icefields are accumulations of snows that do not melt completely. When deep enough, the snow becomes solid ice under its own weight.

Because large masses of ice near the melting point bend and flow slowly in response to pressure, a glacier or icecap follows the terrain when the movement is slow and the direction does not change sharply. Crevasses form in the uppermost brittle layer of the glacier, where there is not enough pressure to permit flow: the ice simply cracks like glass. Crevasses may be very deep at first, but gradually the depths are squeezed together again by the great pressures below. The deepest parts of crevasses "heal" into reconstituted solid ice. Typically, crevasses are limited to about 60 m/190 feet deep, though the ice mass may be 300 m/1000 feet thick.

The layered structure of the ice, visible in the vertical wall of the crevasse, results from yearly accumulations of wind-blown dust and other glacial debris. The sequence and rate of accumulation can be read, much as annual rings chronicle the growth of a tree.

The bare dark patches in the foreground are the dirty surfaces of great blocks of ice. They have broken from the main mass and lie tilted, facing the sun, and are snow-free. Rough light surfaces surrounding the centre are partly melted snow from the previous winter.

8/A JØKULHAUP

This remarkable depression has been carved deep into the glacier ice. A converging pattern of meltwater channels on the surface directs all run-off into the depression, which measures approximately 200 m/650 feet by 50 m/160 feet.

Maps of the icefield display this depression as a lake contained within the ice. Stereoscopic photographs taken from the air at irregular intervals since 1935 show it sometimes full of water, sometimes empty. This close-up view taken from a helicopter on 28 July 1984 shows that the depression had been partly filled but has drained abruptly, leaving behind a broken surface of lake ice and snow.

Abrupt massive draining of lakes contained by glaciers is common in arctic and alpine glacier regions. An ice seal on the bottom may rupture suddenly, and the lake drains rapidly, rather like a giant sink.

The moving water has carved the depression in an unusual swirl pattern. The water probably descends into an active cave system, Castleguard II, that is known to drain into the Big Springs in Castleguard Valley. This basin may be the head of a great cave.

Glaciologists use the Icelandic term *Jøkulhaup* to describe this sudden draining, which is common in the very wet icefields of that island.

The stark beauty of this sterile scene is like a glimpse of the past, when ice and snow covered the whole land.

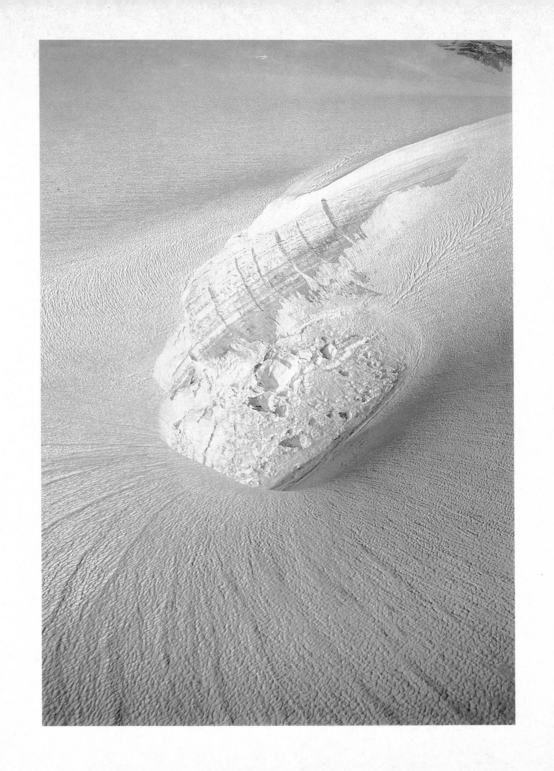

9/SIDE VALLEY GLACIER

Castleguard is bounded on the north side by the Saskatchewan Glacier, which flows from the Columbia Icefield. Along both sides, smaller glaciers join the main ice, like streams emptying into rivers. This tributary glacier had retreated and did not join the main stream when this photo was taken in 1969. In 1984, the tributary glacier had retreated farther, and the Saskatchewan Glacier had sunk noticeably. This is typical of the overall melting and retreat of most glaciers in the Rocky Mountains during recent decades. The figure of the hiker emphasizes the immensity of the distant ice mass, yet it is trivial compared to the Saskatchewan Glacier underfoot.

10/THE UPPER END OF THE CASTLEGUARD MEADOWS

The Saskatchewan Glacier has cut off the upper Meadows and left them "hanging," overlooking the ice. Beside the glacier, the head of the valley is exposed to strong, cold winds, scouring sand and snow. The north-facing slope receives little of the sun's warmth. (The white strips across the foreground are unmelted winter snow.)

The surface of the land is rocky; there is little true soil, no fertile topsoil and very little vegetation. Growing conditions in this bleak landscape are much like those in the Arctic. At Castleguard, which is farther south than Edmonton, altitude, nearby glaciers and prevailing winds have created a landscape like those on the northernmost arctic islands.

Beyond the glacier, small dark stripes slant down the mountainside. These are patches of stunted trees in protected parts of the south-facing slope. The bottoms of these patches were torn off by the Saskatchewan Glacier at the height of the recent Little Ice Age.

11/RECENTLY EXPOSED TERRAIN

A short distance up the sloping sides of the meadow, still within sight of the Saskatchewan Glacier, lie bedrock, thin scattered deposits of gravel and massive boulders, which were rafted in on moving ice. These are vivid evidence that the glaciers once extended much farther down the valley than they do now. Such places are typically barren because there is no soil to support plant life.

12/THE COLD CORNER OF CASTLEGUARD MOUNTAIN

Farther up the northeast face of the mountain, close to the retreating ice, scenery is bleak, new and raw. The mountainside slopes away from the warming sun, and cold winds from the icefield sweep down the moraine. Life has not yet appeared.

It is a lonely place. A person is very small amid the looming ice and the great piles of moraine that it pushed aside only a few decades ago. Yet this menacing ice is only a small lobe of a minor glacier that is spent and dwindling rapidly. The dirty-blue decaying ice and the darker blue-grey limestone rock on all sides present a study in monochrome.

13/ERRATIC AND LICHEN

In glaciated landscapes, it is common to see rocks that are visibly different from the local bedrock. At Castleguard, much of the bedrock is not favourable to growth of primitive plants called lichens. However, some erratic rocks brought here by glaciers may be of a chemical composition that provides a suitable habitat. These two erratic rocks lie side by side. The darker sandstone supports a multiplying colony of orange *Xanthoria elegans*, yet the lighter siltstone remains barren.

Erratic boulders in exposed settings often make good perches for birds. Their nutrient-rich droppings add fertility to the site, and lichens prosper. Well-developed colonies of colourful lichens, such as *Xanthoria,* may be clearly visible at a distance in an otherwise drab landscape.

Lichens are "composite plants," an alga and a fungus growing together. Their tissues integrate symbiotically and produce an organism that resembles neither component. Growth spreads in all directions, resulting in a circular sheet of living tissue several centimetres in diameter. Eventually the central part grows old, dies and falls away, leaving a spreading ring of active growth. The size of the ring is a rough measure of the lichen's age, and can be used to calculate the approximate time since glaciers melted and the landscape could support plant life.

These circular patches are about 3 cm/1¼ inches across. The absence of a central dead zone in most of them indicates that this colony is young and healthy.

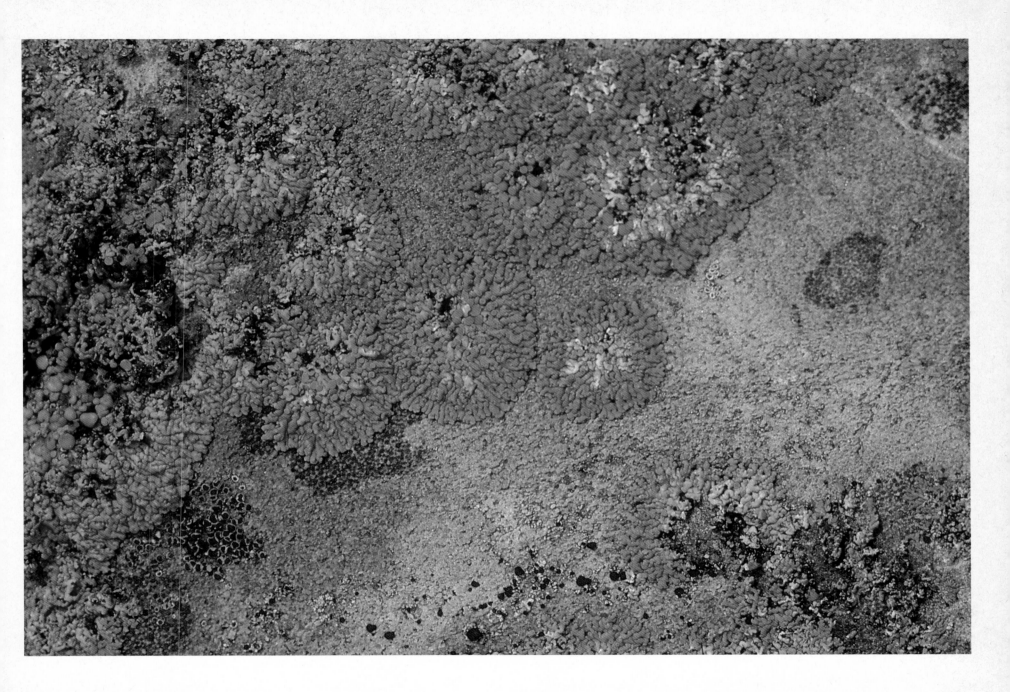

15/LICHEN COLONY

Nearby, map lichen *(Rhizocarpon geographicum)* has established a thriving colony on a different sandstone erratic rock. It too spreads in a circle from the original centre. The black growing edge is the fungus. The algal component establishes itself behind the growing fungal edge and results in the granular yellow composite lichen tissue where reproductive bodies form. The central colony is 5 cm/2 inches in diameter, and has not yet begun to die out at the original growth site.

Sometimes, after the central area of a colony has died and crumbled away, a new colony starts again on the same site. It, too, dies out in the middle and is again recolonized. This repetition produces a target-like pattern.

These adjacent colonies are just beginning to overlap at the edges. Older intersecting colonies that have died out in the central area develop thin, wavy lines that cross frequently and from a distance resemble crude maps.

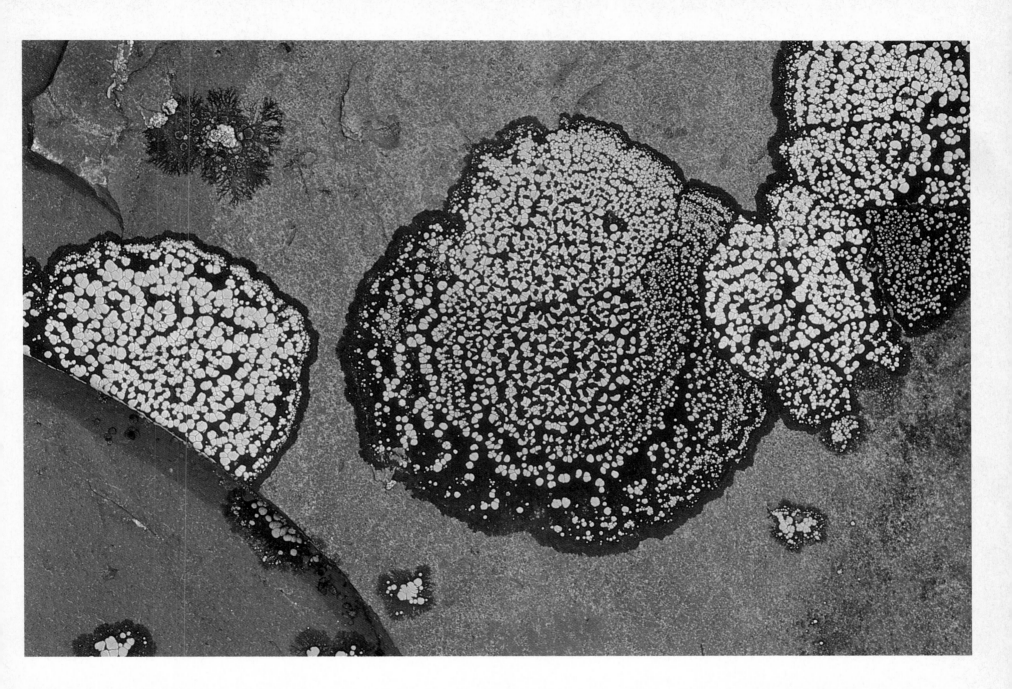

16/PATTERNED GROUND

Fine materials washed down from slopes accumulate in low-lying areas. On terrains with high water content, frost produces "patterned ground." Fine and coarser pieces are sorted and separated. A slow, churning action resulting from daily and seasonal freeze-thaw cycles affects small and large particles differently, moving coarse materials until they meet other large pieces. These soil movements discourage or prevent plant growth.

In many arctic regions, where stronger and more frequent freeze-thaw cycles occur, the patterns may dominate entire landscapes. Eventually, the soil patterns stabilize and developing vegetation grows in patterns that reflect those of the underlying soil. At Castleguard, patterned ground is still rare. Stabilization and revegetation are just beginning.

17/VEGETATION BATTLING FOR SURVIVAL

In an environment that is hostile to life, pioneering plants must find exactly the right niche. Here, in stony terrain with almost no soil, a "frost polygon" pattern has sorted and concentrated enough fine material to provide a small patch of soil for plant life. The frost polygon is about 1 m/3 feet in diameter.

The best "soil" is in the centre of the polygon, where churning movements resulting from freeze-thaw cycles are most active; but here roots would be torn apart by the soil movement. The sterile stones nearby offer no suitable habitat, so plants have started to grow in the only possible place, the outer edge of the soil patch where there is minimal movement but available moisture. The compromise works and plants grow around most of the edge. But they are only a few centimetres high, which indicates that they are only marginally successful. Long-term survival probably depends on increasing soil stability, so that the greater needs of growing plants can be met within this tiny garden plot.

A Succession of Plant Communities

The first plants to become established in a new habitat are pioneers, small, hardy species that can adapt to unfavourable conditions and improve them. Gradually, these plants are replaced by later arrivals, until a plant community develops that is capable of long-term stability.

The higher levels of Castleguard were only recently released from the grip of the glaciers, and plant life here is quite new. As we travel down the valley, we encounter older and more developed plant communities. During a two-hour walk, we experience ten thousand years of plant community development.

18/THE SUMMIT OF CASTLEGUARD MEADOWS

A pass leads from the Saskatchewan Glacier onto the Meadows. The glacier is to the left, below the summit; to the right, the Meadows begin their gentle descent.

At the highest point of the pass there is little real soil. The surface is composed of glacial deposits that have been smoothed and winnowed by frost, water and wind. The exposed mineral surface gives the ground its grey-brown tone.

A few widely spaced plant clusters dot the surface with green. Their random though surprisingly even distribution shows that this is a pioneer plant community. Chance dispersal has initiated growth, and individual plants increase in size and offer local protection and an enriched site for their seeded or cloned offspring.

Such landscape and plant communities must have existed over much of what is now Canada as new land emerged from the last great ice age, perhaps ten thousand years ago. The pattern is strikingly similar to present plant communities on the High Arctic Islands; in both cases the climate did not warm sufficiently to permit development of vigorous plant communities.

19/KRUMMHOLTZ TREES

Still within sight of the Saskatchewan Glacier, but on a warm west-facing slope, higher life forms have taken hold. However, even this more favoured site is still exposed to the effects of the glacier: persistent cold winds, blowing dust and snow and a shortened growing season. These coniferous trees, though very old, are stunted, lack trunks and look more like shrubs. This growth form is called by the German term *Krummholtz*, literally "crooked wood."

The original tree on this site may be very old, long dead and impossible to identify; but lower branches, protected beneath snow and pressed tightly against the ground, took root. Separated from the dying trunk, they formed separate but genetically identical individuals. This "cloning" process does not involve flowers or seed, but permits reproduction under hostile conditions. Similar adaptive life forms are common in the Arctic, and must have been common across Canada at the end of the ice ages.

Mount Columbia is visible above the Columbia Icefield in the background. The dark mass in the upper right-hand corner is the flank of Mount Andromeda.

20/MODIFIED TREE-GROWTH FORMS

Trees that project above the snow-line are very susceptible to "physiological drought." During mild spells in late winter, spring and early summer, strong sun on dazzling white snow can heat dark objects, such as coniferous trees, up to growing temperatures. The warm upper parts begin development and demand moisture that ice-bound roots cannot supply. Trees severely afflicted by these conditions may die from lack of water.

These trees have produced trunks, though they are still visibly affected by damaging winds and particle abrasion. Branches do not grow on the upwind side, and lower branches protected under snow take root and produce clones. Displaying the inherent resourcefulness of life under stress, they also reproduce by growing flowers and seeds.

21/ALPINE FOREST

Near the lower end of the Meadows, growing conditions are more favourable. Plant cover is continuous, and an open alpine forest has developed. The trees here are much higher; they branch evenly and grow faster. However, the tight clumps indicate that propagation is still achieved partly through cloning.

Competition is simple here: a coniferous forest has invaded a tundra-like open meadow. The trees are similar in size and there are few small seedlings, evidence that the initial successful invasion was rapid but that subsequent spread has been limited and very slow.

Throughout the length of the Castleguard Meadows, the pattern of tree shapes, sizes and distribution reflects continental vegetation patterns. The progression of ecological conditions up high mountains in southern regions closely mirrors that encountered when one travels from temperate regions towards the Arctic.

Water Makes Life Possible

Perhaps the most dynamic visual aspect of Castleguard is the presence of water. During the warm season, few places are out of sight and sound of running water. This abundance not only makes possible large areas of rich vegetation, but is the driving force of many landscape processes such as underground cave systems.

22/SNOWBANK VEGETATION

A late-melting snowbank has produced a reliable source of water above a nearly flat area, where rich vegetation and a local meadow have developed. Plant growth has been so successful that enriched soil has developed on hummocks created by the interaction of frost and vegetation. Exactly how this complex interaction works is still not clear, but snow-patch meadows and vegetated hummocks are common in arctic regions.

23/PIONEERING PLANTS BESIDE A GLACIER

Along the west side of the upper Meadows, not far from a glacier and a late-melting snowbank, a reliable water supply has allowed vegetation to thrive in otherwise barren surroundings. Even without soil, moss has established itself on barely projecting small stones and pebbles where percolating water moistens sloping terrain. The narrow gaps between individual plants disappeared and a solid mat of moss developed. Then higher plants, such as sedges and grasses, colonized the moss, strengthened the accumulation and formed a nearly watertight barrier. A pool of water collected, held by the mossy dam. The dead plant material at the bottom turned to acidic peat as a result of partial decomposition; then other acid-loving plants took root and formed an acidic plant community in this chemically alkaline limestone environment.

24/CHAIN OF MELTWATER POOLS

The process of moss colonization and pool formation on a slope can be repeated many times down a watercourse.

Here, a series of pools has formed; each pool receives water from one above and supplies water to one as much as 30 cm/1 foot lower down. Each pool is securely held in place by a peat dam covered with a vigorous community of plants.

The peat dams can withstand periods of drought because peat normally holds enough absorbed water to remain moist for many days before drying out and deteriorating. Such dry periods sometimes occur late in summer if unusually hot, dry weather exhausts the water supplies from late-melting snowbanks.

The dam-and-pool systems at Castleguard are relatively new; the peat accumulations are not deep. However, in Mount Revelstoke National Park, about 130 km/80 miles to the south, a similar series of pools has accumulated peat to a depth of 1.3 m/54 inches. The lowest and oldest layer of peat is 5500 years old. Barring climate change or other external disaster, the dam-and-pool processes can apparently continue indefinitely.

The process has similar origins and development wherever it is found: in other parts of the western mountains, around James and Hudson bays, in Newfoundland and Labrador and among the High Arctic Islands.

25/LOBES OF FLOWING SOIL

Soil-flow or solifluction lobes are common in the soil-mantled slopes in Canada's Arctic and above the tree-line in the mountains. They commonly occur as distinct lobes with steep fronts. At Castleguard, the lobes are a few metres wide, but they may be several kilometres across elsewhere.

Solifluction develops best in arctic and alpine terrains where grades are steep and there are no deep, strong tree roots to hold the soil. The annual spring thaw fills the upper soil with water, making it heavy and movable. Unmelted ground ice in the soil below the thawed, saturated mass facilitates soil flow.

At Castleguard, solifluction has been at work throughout post-glacial time, smoothing down the older moraine ridges. The rate undoubtedly increased during the cooler conditions that prevailed during the Little Ice Age, described in the next section, beginning with photograph 33. But that event has now passed and some lobes in the Meadows appear to be advancing at no more than 1 to 2 cm/½ to 1 inch per year.

26/A WETLAND MEADOW

In a large depressed area of the lower Meadows, the meeting of four surface streams has produced a boggy wetland covered with shrubby growth. The mountain streams enter the wetland separately; because of the low gradient they wind, twist and combine on their way towards a common outlet.

The vegetation appears to be an irregular patchwork of different shades of green. Some of these patches represent different species, but most of them represent clones of single original plants of one species. Plants have individual characteristics, just as people have, say, blue eyes or brown eyes. These tiny characteristics are reproduced in every member of the clone population. Thus a slightly lighter or darker, a more or less vigorous individual would give rise to identical clones. These individuals would be slightly different from a clone of another plant, even one of the same species. The differences are difficult to distinguish at ground level but become apparent from the air, visible proof of the effectiveness of the clone system of reproduction in a harsh environment.

The small brown patches among the green vegetation are sinkholes, part of the underground drainage that connects with caves below. These caves drain much of the Meadows, the bottoms of some glaciers and parts of the nearby Columbia Icefield. The sinkholes are not large enough to swallow all of the water of the wetland, so the excess flows away as surface streams: thus, two independent drainage systems operate on this part of the Meadows.

27/MORAINE

Melting water from snow and ice on Terrace Mountain passes easily down and through the porous brown moraine, but must percolate over the impervious grey limestone beneath. The water reappears as surface streams, which rise at the lower edge of the moraine. The limestone surface was glaciated during the great ice ages, which ended about ten thousand years ago. The moraine lying on top of it is a product of glaciers of the Little Ice Age, which retreated during this century.

28/A GLACIER IN RETREAT

At the lower end of the central Castleguard Glacier, meltwater gushes from decaying ice and cascades over recently exposed bedrock. Only a little moraine material is visible. Nearby are the great moraine piles that mark the farthest advance of the Little Ice Age glacier, but they are no longer in contact with the ice. This glacier stopped advancing early in this century and is now melting rapidly, diminishing in thickness, width and length.

Retreat features such as this melting glacier provide much of the reliable water supply that supports the rich vegetation of the Meadows.

29/WATERFALLS

Throughout Castleguard Meadows, and especially on the high, flat benches along each side, meltwater from the icefield, glaciers and snow pours over ledges. The most picturesque of these is a series of falls over a limestone precipice of the Eldon Formation. The water and rocky debris it carries spread out on the Stephen Formation that forms the valley floor.

The sound and sight of moving water blends with the feel of the wind; the sensations reach the mind as one. This is most fitting, for apart from crustal upheavals that created mountains, everything here is sculpted by water. The water that created icefields and glaciers, that moves rock, carves caves and makes possible all life forms is also the only moving thing in most summer landscapes. Winter stillness is only a pause in its activity.

By late summer in a dry year, most surface snow is gone, and the waterfalls are thin veils, seductively cooling on a hot day. The size of the scoured waterway above and below the falls indicates that these falls have been much larger during spring seasons and during waning years of the Little Ice Age.

30/AN ALLUVIAL FAN

Below the waterfalls, the stream spreads over the valley floor, dropping rocky debris in a huge wedge-shaped alluvial fan. The narrow upper part is still receiving debris from above. It is washed by torrents each year and is almost devoid of vegetation.

The lower, older parts of the fan have stabilized, accumulated soil and organic debris and developed a pioneer plant community. Individual plants, clumps and clones have not yet joined together; but in time they will form a blanket of vegetation.

Pioneer plants and young plant communities reflect the recent origins of neo-glacial landscapes. Even the last ice-age landscapes at this altitude are only marginally suited to vegetation, and a long succession of ever more complex plant communities is needed before a stable community is reached in such harsh environments with such short growing seasons.

31/FLOWERS ON AN ALLUVIAL FAN

High in the mountains, spring arrives late but fall arrives early. During the short summer, plant life develops at breakneck speed and a brief, glorious surge of blossoming appears. Only a few days after the first flowers of a species appear, the rest burst into bloom. Seed follows quickly, because an unusually early frost is always a danger. When early frost strikes, an entire year's seed production may be lost; no new plants will form and continuity of the species will depend on the survival of existing plants.

This early August scene shows the entire vertical sequence at the height of summer. At the top are the sterile mountain peaks. Below lies the central Castleguard Glacier, fed by the Columbia Icefield, with its light-brown lateral moraine. Among trees of the thin alpine forest is the rocky streambed that carries meltwater down to the alluvial fan. The moist, well-vegetated margin of the fan is a floral meadow in full bloom. Two days after the photograph was taken, night-time frost killed the flowers.

This setting is also a dramatic rendering of ecological succession. From barren gravel deposited perhaps a hundred years ago, a series of plant communities has progressed to become a floral meadow of species such as the showy pink legumes of the pea family and the less noticeable cream-coloured *Dryas* of the rose family.

32/A TINY ARCTIC PLANT

Among the disturbed boulders and gravels at the edge of the alluvial fan and in the moraine above is a tiny Dwarf Hawk's-beard, *Crepis nana,* which shows the biological link between the Arctic and the high-mountain regions far to the south. This plant is found only in arctic or high-alpine settings.

Dwarfism is a characteristic of many arctic plants. They are smaller than their southern relatives and grow flat or close to the ground, where it is warmer during the growing season.

The Dwarf Hawk's-beard is a flat rosette of leaves and flowers, 2.5 to 5 cm/1 to 2 inches in diameter, with a long central root that grows through loose surface rock and other barren material to moisture and soil nutrients below. Its family relationship to the common dandelion is evident in the shape and make-up of its flowers; but the fragile, fleshy leaves seem out of place in this harsh setting.

Nestled deep in the crevices and spaces between rocks, however, this plant is safe from trampling, because the sole of a boot would not likely reach the plant itself.

The Little Ice Age

About seven hundred years ago, the world climate cooled slightly. Alpine glaciers moved down valleys they had not occupied for thousands of years. This Little Ice Age, or neo-glacial, event persisted well into this century and its effects at Castleguard lie on all sides, fresh and undisturbed.

33/THE CENTRAL CASTLEGUARD GLACIER

The great ice sheet that covered nearly all of Canada retreated about ten thousand years ago, exposing lowlands that probably revegetated quickly. About seven hundred years ago a minor cooling known as the Little Ice Age took place. Previously retreating glaciers expanded slightly until the present century, when the retreat began once more. Halfway down the Meadows, the central Castleguard Glacier advanced during the Little Ice Age. When the black-and-white photo was taken, in 1918, the edge of the ice had thinned, but was still in contact with its moraine. During the subsequent sixty-six years, the ice deposited much additional moraine, before retreating to the line of sharply defined small dark markings across the centre of the ice. The alluvial fan seen in photograph 30 is visible in the central foreground as a light patch; but the waterfalls that deposited it (photograph 29) are either hidden or were not flowing when the photo was taken.

In the colour photograph, taken in 1984, the moraine piles along each side of the glacier are sharp, new and unvegetated. The lower margin of the ice is thin and much of the exposed bedrock has almost no moraine: the retreat must have been very rapid indeed. The dark band in the foreground is a "drumlin," a glacial deposit with oval margins and a whale-back outline, its higher end upstream. It was exposed during the main retreat of the ice about ten thousand years ago. Its long axis, which runs parallel to the valley walls, mirrors the direction of the ice flow.

34/THE SASKATCHEWAN GLACIER, MORAINE AND TRIM-LINE

During the Little Ice Age the thickening Saskatchewan Glacier overflowed its present course and erased a coniferous forest. The remaining upper forest appears as dark wedges above an abrupt trim-line, which marks the greatest extent of neo-glacial ice.

In the foreground is massive moraine deposited by the retreating ice. There is little fine material or soil for plant life; its nearly barren expanses have been altered only by the movement of water running down from above.

Remnants of tributary glaciers on Mount Athabasca lie in the background, shrunken, retreating and dirty. The melting surface exposes some of the rocky debris the ice is carrying.

35/MORAINE OVERRIDING OLDER SURFACE

The west-facing slopes of the Castleguard Meadows receive the midday and afternoon sun, and are therefore warmer than the east-facing slopes. Little Ice Age glaciers along the west-facing side have retreated farther than those opposite, and can barely be seen from the meadow floor. These glaciers transported much more material than those on the east-facing side of the valley because they erode softer rocks. Their moraine deposits are wider, deeper and thicker. The deposits are relatively new: compare their barren surface to the partly treed land that they have encroached upon, which was itself also glaciated until ten thousand years ago. It is clear that the Little Ice Age advance was smaller than that of the great ice ages, because the foreground and higher background had been glaciated much longer ago.

36/SOUTH MORAINE, CENTRAL CASTLEGUARD GLACIER

The moraine on the south side of the retreating glacier has very sharp ridges. Its sides slope at the greatest angle of stability, given the size and type of material it contains. The cracked, bluish ice is part of the decaying glacier mass, now separated from its major moraine deposits and melting quickly.

The neo-glacial advance has intruded into a *Krummholtz* forest that had been developing slowly since the last great ice sheet departed about ten thousand years ago. Such obvious contrasts between very old and recent events are common throughout Castleguard.

37/LITTLE ICE AGE DAMAGE TO TREES

The south glacier beside the lower meadow advanced considerably during the Little Ice Age, and retreated quickly too, perhaps only eighty years ago. The glacier ploughed into and swept away a mature forest. We can see the limit of the advance very clearly. Boulders in the foreground are the remains of a lateral moraine that, pushed by ice, swept away part of a forest on the near side of the valley. A large dead tree leans against untouched forest. Root damage from the ice and moraine only a few feet away probably killed this former forest giant. Nearby lies the stump of another dead tree whose roots came into direct contact with moving ice and rock.

38/THIN NEO-GLACIAL MORAINE

The south glacier was much larger during the Little Ice Age, and marks of its retreat are visible along the sides and near the toe. A thin moraine of brown sandstone boulders on top of the grey limestone bedrock shows where the edge of the retreating glacier once halted for a few years. Such ice margins are particularly sensitive to climatic change, and the small amount of moraine indicates that the glacier did not stand at this site for long: at this place, the entire cycle of glaciation may have lasted perhaps only a century or so. The huge valley in the background was filled by the glacier at the height of the Little Ice Age.

39/COLOURFUL ROCK OF THE NEO-GLACIAL MORAINE

Erratic rocks were transported by ice from geological formations overlying the Castleguard limestones. Many of these upper, younger formations are brightly coloured, and create sharp visual contrasts when fragments of them are dumped on the uniform grey surfaces of the limestone. The differences of colour reflect differences in the origin and composition of these rocks.

The erratic rock shown here comes from a geologic unit named the "Waterfowl Formation," deposited between 510 and 520 million years ago, during a period known as the Middle-to-Upper Cambrian Interval to historians of the geology of the earth. Rocks of the Waterfowl Formation form an outcrop close to the summit of Castleguard Mountain, where this example probably came from.

The rock is a mixture of silicate silts and sand (yellow and brown), with silty limestones (pale blue-grey). These differing materials were laid down in successions of thin lenses, or layers, with rippled surfaces. The environment at the time of deposition was a very shallow tropical sea floor.

The visible width of this rock is about 20 cm/ 8 inches.

40/SOUTH BENCHES OF CASTLEGUARD MOUNTAIN

Nowhere is evidence of the Little Ice Age advance fresher than on the southern benches of Castleguard Mountain. The scoured-limestone surfaces are strewn with only a thin scattering of glacial debris, indicating a rapid retreat. The thin remaining ice is partly covered with the previous winter's snow.

The recently exposed bedrock in the foreground of the photograph is a veritable treasure trove of intricate small-scale materials and shapes that were produced under the ice. Details of these occurrences are shown in photographs 44 and 94 bottom.

On a cloudless summer day, the benches are private and lonely. In their isolation, all sound is deadened and no echo is returned. The sky, snow, ice, meltwater and dark, sterile bedrock are the essence of solitude, a minimalist pattern of nature.

Under a hot sun, distant benches glitter with meltwater, shimmering in a layer of superheated air, as the cold wind from the icefield ruffles their outline. Simultaneous burning and freezing sensations send conflicting signals to a mind exalted by visual splendour.

A new landscape is emerging; life has yet to arrive.

41/THE SUNNY SIDE OF THE MOUNTAIN

Unlike the Cold Corner in photograph 12, the south benches face the midday and afternoon sun. Though the visual elements – ice, rocks, snow and sky – are the same in both places, the daily rhythm of weather affects the senses very differently. Here, a glacier of the Little Ice Age disappears day by day, exposing its effects on the bedrock below as the ice edge moves back. We are witnessing the final act of an ice age.

The exposed limestone surface in the foreground is so thinly strewn with moraine that it appears like a paved road, abandoned and broken after a few years of disuse. The retreat is so rapid here that each year's melt exposes new stretches of the road.

In the background, avalanche tracks can be seen along the face of the mountain. The vertical meltwater channels in the snow cover of the lower ice are coloured by red algae washed down by the melt. The dark rock to the right of the snow is one of the exposed higher benches, where sunset lingers longest at Castleguard.

42/A CAVE IN GLACIER ICE

We are facing the exit of a natural cave developed at the snout of a glacier on the southern benches. In the natural light, which filters through several metres of glacier ice overhead, the blue is emphasized. Darker stripes and patches are rock debris carried in the moving ice.

This glacier cave occurs in the lee of an obstruction over which ice is flowing. It is only rarely that one can enter such cavities. Being close to the leading edge of the glacier, the ice is now thinned and is quite rigid. The ice roof spans 3 m/10 feet; the abrasion grooves were scraped in it as it flowed over the bedrock.

The sporadic ice motion that creates these caves does not stop once they are formed. Explorers who spend time in ice caves on warm days have observed the roof to shift a centimetre or two very abruptly. This need not be cause for alarm, but the groaning sound that sometimes precedes the shift has sent many a hardy mountaineer fleeing for the outside world!

Glacier-ice caves are transient features. First accessible in 1968, this cave was destroyed in 1971 or 1972 during the glacier's retreat; but there are signs that a similar cave will open again in this glacier within the next few years.

43/GLACIAL ABRASION, POLISH AND STRIATIONS

Glacier ice itself is quite soft. However, sometimes it freezes onto loose rocks, pressing them firmly onto bedrock. The ice then skids the rocks across the bedrock, which becomes gouged and scoured. Both the rock and the surface over which it passes can be ground into fine powder, or "silt." When produced in great abundance at the sole of a glacier where ice slides over bedrock, silt is known as "glacier flour," as it gives melt rivers a floury or milky appearance. Abundant glacier flour can polish the bedrock to a surprisingly high gloss.

Often, grinding, scoring and polishing succeed one another during the waning of a glacier. Here, neo-glacial ice has receded within the past forty years. The ice formerly flowed *up* and around the bedrock shoulder. Many striations have been scored up the rock and over the crest. The whole surface is highly polished. Because limestones are quickly dulled and pitted by dissolution in rain water, we can assume that it must have been exposed quite recently.

In its final decay, the ice here was a stagnant mass. Rocks within it were let down slowly during the melt; some came to rest at rather precarious angles.

White material in recesses of the grey limestone is a sub-glacial precipitate of calcite (see photograph 44). Here, it has been nearly destroyed by abrasion.

44/SUB-GLACIAL CALCITE PRECIPITATES

This bedrock surface lies on the southern limestone benches of Castleguard Mountain. It is only 50 m/162 feet away from a glacier whose recent retreat has exposed this pristine sub-glacial surface. Minute intricate details are better developed, more varied and elegant here than at almost any other known site.

The bedrock was deposited in shallow seawaters, where the original limy silt was raised into patterns of sand ripples. The ripples were buried intact by subsequent deposits. Five hundred million years after they were formed, these ripples are now being exhumed by glacier ice flowing across and eroding them at an angle of about 45° to their crest lines. As a consequence, a pattern of streamlined tiny hills is formed in the rock; the upstream ends are steep, but they taper towards the lee. They are marked by coloured wavy layers of the original ripple pattern, now exposed by the melting of the glacier that ground away the surface and exposed the ripple pattern. The photograph includes about one square metre of the ground.

In the central Castleguard Meadows, retreating glaciers of the last ice age deposited drumlins hundreds of metres long but of precisely this shape (see photograph 33). The tiny scale of the solid rock "drumlinoids" shown here illustrates how, in nature, similar processes can produce similar forms a centimetre or a kilometre long!

Surrounding each micro-drumlinoid and partly burying its tapered lee slope are deposits of white calcite created in the sub-glacial environment. These calcite precipitates (calcium carbonate) are extremely fragile and very soluble; frost and rain quickly destroy them. Deposits as fine as at Castleguard are rare indeed.

45/SUB-GLACIAL CALCITE

This close-up photograph shows some typical features of sub-glacial calcite. The limestone surface at the top is the "stoss," or upstream end, of a micro-drumlinoid. The pressure of ice flowing against it caused local melting and the meltwater dissolved the tiny grooves in the rock. Later, the water froze and the dissolved limestone was re-precipitated as the white, needle-like calcite crystals seen here. These needles are 2 to 3 cm/1 inch in length. They point in the direction of the ice flow.

On the right, recent ice movement has abraded the calcite. In the centre is a tiny bedrock escarpment buried under the calcite deposit.

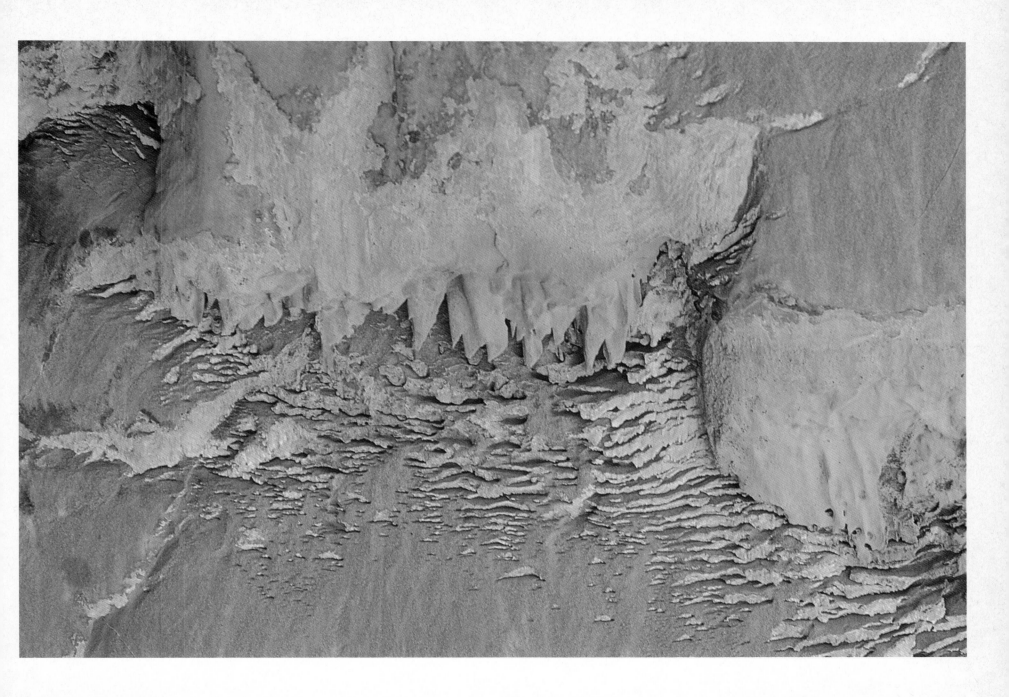

Cave Formation: The Surface

Cave-forming, or "karst," processes begin at the surface of the earth. Slightly acidic water, which contains carbon dioxide from the atmosphere, finds and enlarges surface cracks in limestone or dolomite rock and sinks into them. The effects of water are abundant on the surface rock, where their presence hints at the possibility of caves underground.

46/ALPINE KARST

This bleak scene expresses the spirit of the alpine karst of Castleguard. Mountains of limestone rock and plenty of water from melting snow and ice are the ingredients; time is the catalyst.

In the background, the glacier-hung east face of Mount Andromeda, 3442 m/11,300 feet high, screens our view of the main mass of the Columbia Icefield. In the mid-ground is the "col," or pass, that leads to the Castleguard Meadows; it is a glaciated surface, scoured by flowing ice that receded some ten thousand years ago. An ice-moulded, oval-shaped drumlin, in the distance at the right, tapers downstream (to the left). In the foreground is a limestone pavement with a characteristic pattern of clefts and runnels (enlarged cracks and waterways), created by the dissolution of limestone by rain and melting snow since the ice receded. Water sinking into the clefts in the foreground will next see the light of day at the Big Springs, 6.5 km/4 miles distant and 560 m/1800 feet lower.

There are three basic kinds of cave or "karst" landscape: tropical, temperate and alpine or arctic. This is an outstanding alpine karst scene: it contains modern glaciers as well as formerly glaciated terrain and karst land-forms, on the surface and underground. Most alpine karst lands in the European Alps, the Pyrenees, the Caucasus and the Rocky Mountains in the United States lack modern ice.

The cliff was created by glacier action during the Little Ice Age or neo-glacial period, which reached a maximum during the 1800s. It is 5 m/16 feet high and is part of the Cathedral Formation, Middle Cambrian in age, formed about 550 million years ago.

This limestone is built up in thin layers of lime mud, with fragmented shells and skeletons of extinct sea life deposited in shallow, warm seas. When buried under later deposits, the material became compressed into hard, crystalline limestone with fossils. It contains only a little insoluble material such as sand grains. This is why the individual layers vary only slightly in colour and texture.

Limestone is composed of calcium carbonate ($CaCO_3$), which is quite soluble in fresh water; but this "massive" limestone is very dense and less permeable. Water can penetrate and dissolve it only along minute, hair-line fractures and weaknesses. The horizontal openings divide the layers of rock into separate "beds," which represent breaks in the deposition of the lime mud. When the mud consolidated into rock and was later uplifted by mountain-building pressures within the earth's crust, some slippage occurred along certain bedding planes, which were emphasized. Vertical joints, which cut through the layered structure of the beds, develop in response to pressure when the rock is consolidating. Rectangular patterns are often formed.

Initially, fissures may be no more than 0.001 cm/$\frac{1}{2500}$ inch wide. Water enlarges fissures by dissolving the surrounding rock, and the routes eventually become cave systems. Water entering from the surface also dissolves the rock. This process also forms the channels, grooves, pits and sinkholes that, in turn, provide access routes for more water.

The bedding planes and joints in these limestones are widely spaced, and the restriction of the water to these few lines tends to produce few but large caves.

Even an untrained eye can look at this massive limestone formation, with its grid of fissures, and be fairly confident that it contains caves. A karst geomorphologist would simply say, "Ah! Good rock."

48/WATER MOVES ALONG A BEDDING PLANE

In these thick limestone beds, water has entered and found an easy route downstream, flowing along the horizontal openings in the rock. The water has spent most of its dissolving power higher up and emerges here at the surface as a clear trickle.

Moss is taking advantage of this reliable water supply.

49/ROCKS OF DIFFERENT SOLUBILITIES

This is the face of a vertical joint that was laid bare when glacial action pulled the rock mass apart and carried away the nearer half. Patches of the white, crystalline calcite that originally filled the joint are preserved at the right.

The bedrock shown here belongs to the Pika Formation, the topmost layer of cave-forming rock at Castleguard. It comprises grey crystalline limestone with interlayers of a buff-coloured, silty dolomite.

Dolomite is a double carbonate of calcium and magnesium ($CaMg(CO_3)_2$). Dolomite is soluble and it is considered to be karstic. However, pure limestone, composed of the single carbonate mineral, calcite ($CaCO_3$), is more soluble, hence more karstic.

On fine days a film of snow meltwater drains down the rock face. The water is delicately etching out the more soluble limestone, leaving the less soluble dolomite layers protruding.

50/DISSOLVING AND DEPOSITING LIMESTONE

High among the south benches, meltwater from a deep snow pack has penetrated a bedding plane in the limestone underneath, and has dissolved the rock at a temperature only a little above the melting point. Now it has seeped out into the sunlight. The water is warming rapidly and evaporating, depositing some of its calcite as it does so. The whitish rimes of calcite are soft and will soon be destroyed by rain, or by melting snow next year.

51/SOLUTION PITS AND CHANNELS

Large amounts of surface water, slightly acidic as a result of dissolved carbon dioxide, encountered cracks and joints in surface limestone and enlarged them, gaining access to an underground drainage system.

A smaller and later trickle of acidic surface water has carved its own barely visible channel on top of the rock. Curving downward, the water encountered the long, narrow sinkhole and was swallowed up. The surface channel is partially blocked in two places by minute amounts of rocky debris.

52/RINNENKARREN

Very small solutional channels, grooves and pits are known collectively by the German term *karren.* Well-rounded channels carved into a larger vertical face are *rinnenkarren,* or runnels. The water that created them drained through the mixture of soil and rubble in the background before becoming focussed in closely spaced but separate courses.

The runnels are cut into the sides of clefts, which are solutionally enlarged vertical joints. They are usually straight and deep; this cleft in upper Cathedral limestones in the upper Meadows is 4 m/ 13 feet deep.

53/KLUFTKARREN

Clefts, or *Kluftkarren,* are the longest, straightest and deepest of the small-scale solutional land-forms. Examples here are up to 20 cm/8 inches wide. They extend downwards 1.5 m/5 feet to a partly opened horizontal bedding plane that receives their waters.

Joints that intersect at right angles divide the rock into regular blocks. In Yorkshire, the clefts are called "grikes" and the blocks "clints." Together they form "clint-and-grike" topography, a term that has come into international usage.

This example is in upper Cathedral limestones at the north end of Castleguard Meadows at an altitude of 2350 m/7770 feet. It has probably taken local snow melt and rain ten thousand years to widen these joints and create unsupported corners and shoulders of rock, which develop microscopic cracks. Once solution enlarges them, frost can penetrate. At the junction in the centre of the picture the corner is frost-cracked and being wedged outwards. A covering of vegetation and soil tends to prevent this kind of degradation by furnishing insulation.

54/CLINTS AND GRIKES MODIFIED BY FROST

These clints and grikes have been subjected to extensive frost action. All the surfaces of the clints, or blocks, have been frost-shattered. The tops of the two blocks in the foreground have been reduced by frost to fine rock chips which eventually accumulate in the grikes, or clefts, to form soil. Pioneering plant life has become established in one grike in the midst of the rocks.

The foreground clints have been most affected by frost because they stand highest, lost their protective soil cover earliest and have been exposed to frost action the longest.

In the upper left-hand corner, thin soil formerly protected the original grike pattern and preserved conditions like those in the photograph 53. A small remnant of the former soil cover is still visible in the upper right-hand corner, but appears to have been mostly washed away in recent times.

55/LIMESTONE PAVEMENT WITH ENCROACHING ALPINE VEGETATION

Joints enlarged by water create large patterns that have the appearance of flagstone pavements. Some are as regular as pavement in a courtyard; others look like "crazy paving." The pattern depends on the rock joints.

Limestone and dolomite pavements are widespread in Canada. Well-developed examples can be seen along the St. Lawrence between Quebec and Montreal, around Ottawa, on the Niagara Escarpment and its extension on Manitoulin Island and in the Interlakes country of Manitoba. These pavements are in innumerable patterns, but rarely are they as sharply defined as this one.

Here, two nearly rectangular joint patterns intersect at an acute angle. Solution has opened up some parts and left others untouched, creating a complex, discontinuous pattern.

Only a few hundred metres from the barren clint in photograph 54, this site is more sheltered. (It is 100 m/325 feet lower.) Small alpine plants have established themselves in the scarce soil of the solution clefts. They trap wind-borne dust, which supplies more soil and permits the smallest plants to spread. A large central clump of the plant *Dryas* sp. has extended over small adjoining clints. A few feet away are four small cushions of pink flowered Moss Campion, *Silene acaulis*. In the background the spreading process is far advanced and the surface is almost completely covered by vegetation.

Once well-established in solution clefts, vegetation and soil bacteria produce large amounts of carbon dioxide, CO_2. This dissolves into downward-draining water and makes a weak solution of carbonic acid (H_2CO_3). The capacity of surface water to dissolve limestone may then be enhanced as much as sixfold. As a consequence, the spread of vegetation across a previously bare pavement usually implies that its rate of development will increase and that, in particular, solutional pits and runnels will begin to form on the clint surfaces.

56/MOSS-CAMPION

This arctic and alpine species, *Silene acaulis*, grows well at Castleguard; it is found at all elevations. It forms small cushions with pink flowers faintly visible at several places a few feet away from the central cluster of plants in photograph 55.

At Castleguard, the average annual temperature is less than 0°C/32°F. Overnight freezing may occur more than a hundred times each year, and permanent ice appears in soils above 2100 m/7000 feet.

This well-vegetated solutional pavement in massive limestone of the Eldon Formation is on a sheltered site at approximately 2200 m/7218 feet.

Most of it is insulated by vegetation and shows little frost damage. The patch in the foreground and the block upon it is of the same limestone, but has no plant life to protect it. Frost has split the exposed block and reduced its upper surface to rubble. Ice, which accumulates in the rubble from time to time, is wedging the two halves apart.

58/FROST-SHATTERED ERRATIC BLOCK

The brown boulder of silt-stone was brought here in glacier ice. It is called an "erratic" because it is unlike the local bedrock and so must have originated elsewhere. It has been shattered by exposure to repeated freeze-thaw cycles. Wind, gravity and snow melting down the slope have scattered broken pieces over several metres. Part of the original rock is still in place, but it has been completely split into flakes.

59/FROST-SHATTERED LIMESTONE PAVEMENT

This site of Eldon limestones is at an elevation of 2200 m/7200 feet. It is more exposed than the nearby site in photograph 57, and here the clint, or block, has been shattered and dislocated by frost action.

Recent shattering has broken up an earlier pattern of narrow solution grooves on the clint. The shattering may be associated with the slight but general cooling accompanying the neo-glacial ice advances of the past seven hundred years.

60/A SEA OF ROCKS

At 2500 m/8200 feet on the southern shoulders of Castleguard Mountain, it is bleak indeed. A fresh lateral moraine in the background marks the local limit of neo-glacial ice advance. At the left is a northwest peak of Mount Bryce.

After the main glacier ice receded from these high limestone and dolomite benches and ridges more than eight thousand years ago, solution began to create pavements in the scoured surfaces. But as soon as cracks were opened slightly, frost action ruptured the rock to create this *felsenmeer*, or "sea of rocks." A surface of angular, shattered rubble covers bedrock to a depth of 30 to 50 cm/12 to 18 inches. Solution and frost continue to attack the bedrock but, as the veneer deepens, the frost destruction diminishes.

Here, at the top of the Pika limestone formation, an abundance of frost rubble comprises much of the surprisingly large moraine ridges of the neo-glacial period. It is easy to visualize how a comparatively small glacier could move such loose and broken rock.

The average annual temperature here is −6° to −7°C/20° to 22°F. The bedrock should be permafrozen and impermeable to water. However, there is one major sinkhole in the *felsenmeer*. "Frost Pot" is the long, narrow shaft at the centre of the picture. There is rubble around its entry, but water can be seen flowing into it.

This shaft is vertical. Cavers descended to a depth of 30 m/97 feet, where the shaft becomes very constricted. Coloured dyes mixed into the sinking waters there have been recovered as little as three and a half hours later at the Big Springs, 4 km/2½ miles distant and 750 m/2500 feet lower. This is very fast, and suggests that the water is falling down a long series of vertical shafts.

This efficient system of open shafts in a permafrost climate is maintained by water from the large snowbank at the top of the photograph. The snowbank faces northeast on a lee slope of the prevailing westerly winds. Its small melt stream is heated by the hot summer sun to as much as 12°C/54°F when entering the Frost Pot. The water warms the shafts, producing a cylinder of warmed rock descending through the permafrost.

61 / ROCKS BURYING
   A LIMESTONE
   PAVEMENT

This picture illustrates the competition between karst solution, frost and solifluction processes that are found at high altitude at Castleguard. The bedrock is of Eldon limestone, part of a sunny, south-facing bench at 2380 m/7800 feet. The limestone pavement was created after the last ice age. It is now being buried by the advancing lobe of moving rocky debris. The debris, or "talus," has accumulated at the foot of a low cliff. During the Little Ice Age, some ice formed in the talus pile and it began to move downhill. Once buried by talus, the limestone pavement will be protected from further attack by acidic surface water as the talus absorbs and neutralizes it.

62/A SINKHOLE SWALLOWS SOIL

In photograph 61, a lobe of rock debris is flowing down a slope to bury solutional fissures that help form caves. Here the process is reversed, and fissures are consuming the soil in a previously vegetation-covered area.

In the tundra at the upper end of Castleguard Meadows, 20 to 40 cm/8 to 16 inches of clay and silt cover an ice-scoured limestone surface; the silt probably obstructed bedrock drainage of the type that creates caves. The clay and silt soil was once covered by the rich, hummocked vegetation visible in the background. But decaying plant matter produced carbon dioxide, which made the surface water percolating downwards towards the bedrock acidic. It attacked the bedrock, widening the grikes or fissures until the unsupported soil collapsed into them. Gradually the water created a functional drain through the rock.

The rush of water during spring run-off washes soil away from roots, breaks up the mat of vegetation nearby and washes soil and plants underground. Eventually, all nearby soils and vegetation will disappear and a major sinkhole will develop.

63/A DEEPLY BURIED SINKHOLE

Sometimes soil erodes by washing away through a natural pipe within the soil itself. The process, called "piping" or "suffosion," is particularly well-developed in areas where there are caves because solutional shafts in the underlying bedrock encourage the soil pipes to drain into them.

This outstanding example of karst suffosion is in the upper Castleguard Meadows, at 2240 m/7300 feet. Well-vegetated glacial "till" 3 m/10 feet thick lies on top of limestones of the middle Cathedral Formation. The till is mostly silt and clay, but also contains some bigger stones. The surface has been transformed by vegetation and patches of ice growing in the soil into a pattern of earth hummocks, which are common throughout arctic and alpine Canada.

A stream has eroded the till cover and opened a sinkhole into bedrock 30 m/100 feet away. A pipe in the soil has developed to connect to it. Silt and clay is washed through the pipe during the spring thaw and rain storms, and a funnel-shaped sinkhole has been created. The pebbles, cobbles and boulders in the till are too large to be carried by the streams and have been left behind as a "lag," or layer of rocks, leading to the pipe.

64/ A GLACIALLY ERODED SINKHOLE

This sinkhole is at 2340 m/7700 feet, at the north end of Castleguard Meadows. Mount Andromeda, on the far side of the Saskatchewan Glacier, is seen in the background.

During the Little Ice Age a thick and powerful glacier advanced across this massive limestone surface in the upper Cathedral Formation. It has now receded 250 m/820 feet and is out of view to the left. The glacier overran an elliptical sinkhole that had developed in the major joint extending across the scene from the lower left.

The flowing ice quarried the upstream – left-hand – wall of the sinkhole. The right-hand wall facing the oncoming ice was abraded and rounded. Sub-glacial calcite precipitates (as in photographs 44 and 45) are deposited liberally upon the abraded surface.

The resulting highly asymmetric land-form permits us to gauge the amount of sub-glacial erosion that occurred during several hundred years of direct neo-glacial action. Although its shape is greatly changed, the sinkhole survives and efficiently supplies water to the underground cave system, as it probably did while buried under the glacier. A large stream can be heard flowing within the glacier, but no water emerges. It is probably plunging down another sink like this one.

Fellow geomorphologists will appreciate that this form is the precise geometric converse of the *roche moutonnée*, a type of glacially streamlined hill.

65/SINKING STREAM

The meltwater from snow-fields north of Terrace Mountain on the east side of the Meadows supplies a summer stream that sinks here, at 2290 m/ 7500 feet, the very top of the Cathedral limestone. The water returns to the surface at the Big Springs, 550 m/1800 feet lower down.

This is an example of a young sinking site. The water is descending through narrow clefts opened in the channel bed. A small increase in the volume of flow will swamp the clefts, and the excess water runs as much as 400 m/1312 feet down the stream channel to sink in older, lower clefts.

Cave Formation: Underground

Karst, or solutional processes, produce their most interesting effects underground: they erode the rock, enlarging small joint and bedding-plane spaces into large, complex systems of passages.

At Castleguard, at least three systems of passages have been produced. The topmost has been abandoned by the water that created it. Two lower, newer systems of passages now carry away the water, except during hot summer weather when excess meltwater backs up into the upper cave, sealing its entrance passages.

At the downstream end of Castleguard Cave is the only entrance accessible to explorers. It is located on the north flank of the valley of Castleguard River, just below the edge of the Meadows.

In summer, the temperature of the entrance zone of Castleguard Cave is at the freezing point, and it is wet everywhere. Beyond the caver and the fire is the limestone pavement over which floodwaters pour, keeping it clear of all debris.

The entrance to the cave was probably discovered in 1919 or 1920. Pack-train-tour operators began to guide groups of visitors to the Meadows in the mid 1920s. They would halt here in the afternoon in the hope that one of the frequent and spectacular meltwater floods would burst from the cave.

The cave has now been explored and mapped for a distance of 18 km/10.8 miles to points at which it is blocked by glacier ice. It passes beneath Castleguard Mountain and branches underneath the southeast sector of the Columbia Icefield. The farthest passages are 380 m/1200 feet higher than the entrance. Explorers ascend inclined bedding planes and joint fissures rather steadily for nearly all of the distance.

The cave may be divided into three parts. First is the downstream or floodwater zone. This becomes an ice cave in winter. Next is the central cave, mostly a single passage that passes beneath Castleguard Mountain. The ground-water river that created it now flows through a lower cave, Castleguard II, which is inaccessible to cavers. The central cave is hundreds of thousands of years old and contains sparkling grottoes.

The third section of the cave is the headward complex, a warren of smaller passages and pits in the rock beneath the icefield. Some of these carry much water in the summer, though they are dry in the late winter. Other passages are abandoned, like the central cave. Approximately 2000 m/6600 feet of new galleries were mapped there by expeditions in 1983 and 1984, and exploration continues.

Castleguard Cave has many hazards and can be attempted only by experienced, well-equipped explorers. Access is controlled by Parks Canada; applications should be made to the Superintendent of Banff National Park.

67/THE ENTRANCE PASSAGE

Cave passages are of three basic shapes. There are excellent examples of all three in Castleguard Cave.

This chamber, 50 m/162 feet inside the cave, is an example of the breakdown shape and is typical of the first 1500 m/4900 feet of the cave, where flooding and freezing processes predominate. The chamber resembles the interior of a mine, and is the result of rock slabs falling from the ceiling and walls. Summer floodwaters burst from the low passage at the rear and spread out across the floor. They subject the walls on both sides to dissolution, hydraulic pressure and abrasion by pebbles that are swept along in the water. The undermined walls weaken the ceiling and contribute to its piecemeal collapse.

This photograph was taken in early July, just before the first flood of the year. The air temperature is −2.5°C/27°F. The remains of the previous summer's floodwaters lie frozen on the floor. The few scattered flakes of rock detached by frost during the freeze-up mark the amount of erosion during the past winter.

68/THE ENTRANCE
COLD ZONE

The first few hundred metres of the cave are the first to be frozen and last to be thawed.

This picture was taken in April. A pond 80 m/ 260 feet inside the cave has formed from residual floodwaters; it is about 60 cm/24 inches deep. In the winter cold it has frozen slowly and steadily from the surface downwards, heaving up domes that crack open. Consequently, the ice is as clear as the finest glass and displays clean and simple geometric lines. There are none of the opaque layers and patches produced on surface ice by fallen snow or rain and snap thaws.

The winter air is so dry that the ice surface forms a good mirror. There is no moisture on the rock surfaces here, although deeper in the cave (see photograph 74) they are permanently damp. Inescapable, bone-chilling draughts blow through Castleguard Cave except when the entrance passages are sealed off by floodwaters. The cave is long and passes through the heart of a mountain, where it is well-insulated. Winter air temperatures outside are very low, but inside, geothermal heat from below raises rock and air temperatures to as much as 3.5°C/38°F in the middle of the cave. Air here is warmer than the air outside, and flows up the passages to the base of the icefield. Cold outside air is drawn in through the entrance to replace it. Outside air is coldest in the middle of the night, so the draught in the cave is always strongest then.

Cold air begins to flow inwards in late September or October. As the winter progresses, a freezing front moves steadily inwards, until late March or April. The direction of flow then reverses. Cool, damp air flows from the interior to the warmer exterior. This action slowly thaws the cave from the inside out.

69/BUBBLES IN CAVE ICE

Caves are places of natural extremes. They may be swept by terrifying and inescapable floods; they may exist in unearthly stillness and total darkness without change of temperature or humidity for thousands of years.

Such violence and stability have combined to create this scene near the mouth of the cave. Remnants of summer floodwaters trapped in pools have been frozen steadily throughout the winter. Air migrates during the freeze-up and becomes trapped as bubbles of various shapes, up to a few centimetres in diameter. The patterns they form in the ice are almost infinite in variety. The double ellipsoidal row of larger bubbles is about 30 cm/ 12 inches in length.

70/ICE COLUMN

Local seepage waters are able to penetrate the entrance cold zone in only a few places. The flow evidently continues well into the winter season; it is frozen immediately upon entering the principal passage. This column and the ice in the preceding photo are ephemeral: they are created anew each winter; floods destroy them every summer.

71/ABLATING ICE COLUMN

It is the end of the winter. The water supply to the column was halted by freezing long ago. Very cold, dry air blowing in from the cave entrance is evaporating the ice, perforating it and eroding it to knife edges. The upwind cave walls have dried out so completely that dust can be blown off them and deposited on the ice. It is a cold desert scene.

72/"P8," THE 26-FOOT POT

This obstacle 100 m/330 feet inside the cave halted early explorers. It is a vertical shaft, 8 m/26 feet deep, which has developed on a small fault in the rock. For scale, the pencil-thin rungs of the cavers' aluminium ladder are 25 cm/10 inches apart.

So far as is known, this drop was undescended until a party of karst scientists and cave explorers based at McMaster University (Hamilton, Ontario) began systematic explorations in the summer of 1967. The two leading explorers were briefly trapped by a rise of floodwaters during an early visit and, with their rescuers, narrowly escaped being caught for many days. Since that time all explorations have been made during the winter, usually in April. There have been ten major expeditions since March 1968.

The cave entrance passage is a horizontal gallery ending at the top of the ladder. Behind the camera lies the cave interior. The shaft was not created by a waterfall tumbling down, as most natural shafts are, but by floodwaters flowing straight *up* it under hydrostatic pressure. Such caves, formed in a completely water-filled passage under hydrostatic pressure, are described as "phreatic."

The shaft is highly abnormal, perhaps unique, because it has been shaped by a combination of freezing and flooding. For most of the year it is not filled with water. In winter, cold air flows in and chills this shaft to −5°C/22°F or lower. Water forced into cracks by earlier flood pressure freezes, rupturing the rock. The face of the shaft seen here is being shattered everywhere by thaw-freeze action. The large slab seen in the foreground is exceptional; it is the only big rock to fall in seventeen years of observation.

The smaller rock fragments accumulate at the foot of the shaft. Most are small cubical blocks. When the floods come, bursts of faster flow are interspersed with quieter periods. Fast flow sweeps the rocks up the slope. They roll or slide back down during the next quiet spell. The flood mimics the actions of wave break and backwash on an ocean beach, and the rock fragments are smoothed like beach shingle. The dark-grey, rounded and highly polished pebbles in the foreground were produced by this flood action.

When the floods are at their peak, fragments smaller than about 6 cm/2½ inches in length can be lifted straight up the shaft. Its walls are marked by white scars where they have been struck by such pebbles. From the top of the shaft a pebble bar extends 5 m/16 feet along the passage towards the cave mouth. During flooding some of its pebbles are swept out of the cave and replenished by others lifted from below.

73/THE ICE CRAWL

For 200 m/650 feet beyond the bottom of P8 shaft the cave passage is horizontal. Residual flood-waters are trapped here and create a low-roofed canal. As the winter progresses the water freezes and the passage necessitates a long icy crawl, negotiated partly on hands and knees, partly on the stomach. It is a part of the cave that explorers never forget!

74/"P24," THE 80-FOOT POT

We are 2 km/1.2 miles inside Castleguard Cave, at the start of its central insulated section. Air temperatures are always a little above freezing here.

The ground-water river that created the central cave now runs through a lower system. These abandoned or relict water passages carry occasional small streams and dripping or trickling seepage waters that may deposit stalactites.

P24 is a shaft 24 m/80 feet deep and connects the central part of the cave to the entrance zone. It is illuminated here to its full extent. The climber, who can barely be seen on the ladder, is 10 m/33 feet above the floor.

Like P8, this is a phreatic shaft created by ground-waters flowing straight up. The right-hand wall is shallowly indented and smoothed. This is characteristic of cave passages that have been enlarged by dissolution in a complete and permanent fill of water flowing under hydrostatic pressure.

The elliptical shaft is a widening of a small vertical rock fault close to the ladder. Some "breccia" (rock shattered by movement of the fault walls and later cemented by calcite) gives the left-hand wall a rough, angular appearance. White, broken traceries in the lower right-hand wall are calcite veins. (The calcite fills smaller fractures.)

75/THE SUBWAY

This passage is known as "the Subway." It is now abandoned and nearly dry. It originated at the junction of a horizontal bedding plane and a line of small, vertical joints. Water flowing rapidly under hydrostatic pressure enlarged the original opening radially outwards. The Subway is straight as a die for 500 m/1600 feet, on a 5° decline; the bedding plane dips towards the foreground. Water formerly moved through the Subway directly to the base of P24 shaft, where it was forced upwards to the entrance zone of the cave.

This passage is an internationally celebrated example of the simplest and most nearly perfect phreatic shape. The most efficient cross-section for a pipe is the circle, as in our domestic water pipes. Nature strives to obtain the circular form as well, but rarely succeeds because of lack of uniformity in the surrounding material.

The lower half of the Subway contains deposits of clay, which settled out of the last waters to fill the passage. A small trench has been carved in the bedrock floor by a stream that has invaded in the comparatively recent past.

76/THE START OF
FIRST FISSURE

To the cave explorer who is "psyching up" for a new winter expedition into Castleguard Cave, the great obstacles to be overcome are at the First and Second Fissures. These are two narrow and deep stream canyons, both very long and unavoidable. In most places the canyon floor is too narrow to walk in: progress is made by a seemingly endless sequence of traverses along narrow, sloping ledges or on blocks of wedged fallen rock called "breakdown," climbing up or down every few metres to a less exhausting route.

This scene is at the very start of the fissures. A narrow canyon has been created by water eroding downwards beneath an original small phreatic gallery, preserved as a lustrous roof. The canyon has long been abandoned by the river that carved it and a part of its left-hand wall has fallen away along a prominent vertical joint.

77/VADOSE CANYON

The First and Second Fissures of Castleguard Cave are excellent examples of the "vadose" form, one of the three basic cave shapes. "Vadose" means "above the water table," and refers to passages that contain air while they are being formed. ("Phreatic" passages, one recalls, are water-filled during formation.)

In vadose passages, water flows according to gravity, because there is air above the water as in ordinary surface rivers. Therefore, the water erodes only downwards and sideways. The simplest vadose passages are narrow canyons but they can be widened when the action of the water undermines the walls. Simple canyons develop best where the gradient is steep. Castleguard's 5° to 6° grade is sufficient.

The walls of vadose passages are rough. (Smoother surfaces are created by slower-flowing phreatic waters.) The wall surfaces project where the rock is less soluble.

Throughout most of their length, the fissures are 10 to 20 m/33 to 66 feet high. Here we see only the lowest few metres of a sinuous 20 m/66 foot canyon.

78/"HOLES-IN-THE-FLOOR"

Like the Subway (photograph 75) downstream from First Fissure, Holes-in-the-Floor, shown here, was formed by water passing through it under hydrostatic pressure. Flowing downstream from the Second Fissure, the water was forced gently downwards (downdip) through it before turning to ascend through the Central Grottoes. Holes-in-the-Floor is 700 m/2300 feet in length, and 4 to 5 m/13 to 16 feet in diameter.

The passage is not really as circular as first impressions suggest. After it was drained, a considerable stream flowed along the floor where it carved a trench 12 m/39 feet deep. (It can be seen at the bottom of this picture and in the centre of photograph 84.) Together the passage and trench create a shape like that of an old-fashioned keyhole, a barrel over a deep, narrow stem.

However, the passage appears to be round because, after the trench was cut, the entire passage filled again with slowly flowing water, which deposited layers of clay over the stem. At least three layers can be seen, suggesting that the clay was deposited during several floodings, separated by intervals when stalactites were deposited. The clay filled the passage to within one or two metres of the ceiling, but it has been partially removed by streams flowing in the trench beneath. These undermine the clay until chunks collapse abruptly into the water. The bridges of clay remaining between the collapsed sections can be seen clearly.

"Holes-in-the-Floor" was named by the cavers who dash from clay bridge to clay bridge along steeply sloping walls slick with mud, hoping that momentum will carry them forward to the next bridge before gravity takes them down into the trench!

79/THE WATERFALL ROOM

At several places in the Fissures, modern streams enter through overhead shafts. They flow along the fissure floors and then sink away into impenetrable slots leading to Castleguard II below. These "invasion waters," although thunderous showers in the summer, are much smaller than the river that carved the Fissures.

Where they enter, the invasion streams have carved platforms or niches that broaden the Fissures and provide welcome resting points for cavers. The lower caver is propped in traversing posture in a typical fissure; the higher caver stands easily on a broad ledge, where pools of drinking water can be found even at the end of the winter. The ledge is decorated with banks of creamy calcite flowstone. The invasion streams once deposited them, but are now eroding them. The rock in the foreground shows the rough, fresh appearance created by dissolution under waterfalls.

80/THE CENTRAL GROTTOES

Separating First Fissure from Second Fissure are the Central Grottoes. They were created when a ground-water river draining from Second Fissure passed below the water table and encountered a series of faults. The river climbed gently up through what became the grottoes for 800 m/2600 feet before spilling into the head of First Fissure.

The grottoes offer the most diversified erosional scenery in the cave, as well as its loveliest deposits. The passage switches between faults, dikes and bedding planes frequently, and the shape and size of the grottoes change with each shift.

The passage follows one of the sedimentary dikes that are common in walls and ceilings in this part of the cave. (It can be seen at the upper right.) Dikes are great cracks in the original rock that filled with sand and became cemented later by limestone deposited from solution.

Abundant clay was deposited when this passage was last filled with water; as it drained, streams combed down the clay to create the regular banks. There is no stream flow today, but seepage waters are depositing stalactites.

81/CATHEDRAL
PASSAGE

The vaulted form of this passage began with a vertical fracture, which was enlarged upwards by dissolution.

Some pure-white calcite deposits are seen in the ceiling. Water enters here throughout the winter, supplying the pool to the left.

82/BEDDING-PLANE PASSAGE IN THE CENTRAL GROTTOES

This passage was created by water dissolving the rock upwards from an initial penetration along a horizontal slot. The slot is now buried under clay left behind by receding water. The water combed clay down towards a shallow central trough. Then trickling water carved a tiny trench along the base of the larger trough. Drips from the ceiling punched out the splash pits in the clay on the right. The walls and ceiling of the passage are honeycombed with pockets where water has eroded somewhat more soluble rock.

This is the most common type of phreatic gallery in limestone caves. It is wider than it is high because water easily erodes the horizontal slot. The more efficient circular shape seen in the Subway (photograph 75) is rare.

83/GROTTO

The accompanying scene conveys the spirit of the Central Grottoes. The passage has a phreatic form. Clay deposits cover the floor. On the right the clay has been deeply eroded; we can see that it has been laid down in layers. Stalagmites and crusts of calcite cover the floor, while stalactites and more exotic deposits sprout from the ceiling.

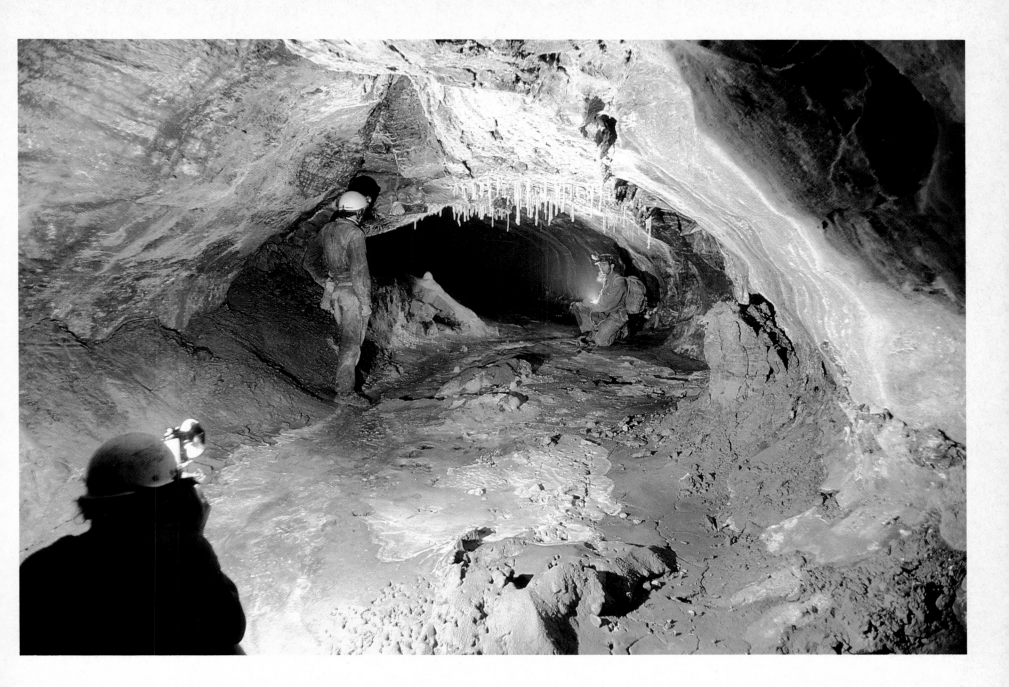

84/ASYMMETRIC GROTTO

A horizontal bedding-plane passage (such as in photograph 82) widened until it met a sandstone dike in the rock, on the left; a vaulted recess was then opened beside the dike. The rough texture of its sandstone contrasts strongly with the lustrous, smoothly eroded surface of the limestone.

Banks of combed-down clay cover the floor and the lower walls. In the right foreground are crusts of calcite so delicate that a touch can break them. The larger stalactites (which hang from the ceiling), stalagmites (which are built up from the floor) and draperies in the centre were deposited by waters seeping down between the limestone and the dike. The longest stalactite is 160 cm/63 inches long.

85/STALACTITES

Stalactites of differing sizes and shapes are abundant throughout the Central Grottoes. The following series of photographs illustrates their variety in more detail.

86/SODA-STRAW AND CARROT STALACTITES

Ground water seeping into caves may become overloaded (or supersaturated) with dissolved limestone. When this happens, it will precipitate part of the load and deposit it on the cave rock. The process usually takes place where a thread of water, flowing under hydrostatic pressure through tiny cracks, suddenly enters a big open space such as a cave passage. As the pressure in the water is released, carbon dioxide in the solution may diffuse into the atmosphere or some of the water may evaporate.

Deposits caused by rapid evaporation are loosely bonded, dull and earthy. They are common in niches on cliffs and beneath old stone bridges, as well as in cave entrances.

Deposits created by the slow diffusion of carbon dioxide from the water yields tightly bonded, regular crystals of calcite that are translucent or brilliant. These deposits are usually found in three common forms: stalactites, stalagmites and flowstones.

Stalactites grow downwards from the ceiling at points where water seeps through. The soda straws seen here are up to 50 cm/20 inches in length.

Very often the central canal of the soda straw becomes obstructed by calcite. Feed water then flows down the outside of the stalactite, depositing successive sheaths of calcite. The result is fatter, more robust stalactites that look like carrots. There are two examples in this picture, one large and well-developed and one just forming.

87/SODA-STRAW STALACTITES

Caving demands physical stamina and mental discipline. Yet among the challenges the caver must overcome, there are rewards: objects and scenes of overwhelming fragile beauty. The sight of these exquisitely delicate soda-straw stalactites is one of the great rewards of cave exploration.

A soda-straw stalactite is a single layer of crystal that encloses a central canal. Feed water descends the canal and deposits calcite at the tip before dripping off the end. It seems incredible that their own weight or earthquakes have not snapped them long ago; but these are nearing completion, when the upper stalactite will join with its lower, sturdier counterpart, the stalagmite. If this occurs, the upper portion will be supported and the anchored whole will be more rigid and less vulnerable.

88/WIND-FLAGGED HELICTITES

In many caves soda-straw or slender carrot stalactites curve into spirals or sprout protuberances that grow in all directions. Such features are known as erratic stalactites or *helictites*, from the Greek word for "wandering."

In Castleguard Cave the most important cause of erratic growth is the continuous draught from one direction passing across the growing deposit, which causes horizontal growth. We speak of these stalactites as being "wind-flagged." The horizontal portion or "flag" points downwind. Some flags are as much as 30 cm/12 inches long, and probably have taken thousands of years to grow to that length.

The summer airflow predominates, travelling down the passages to the southeast and out through the explorers' entrance. This has caused hundreds of stalactites, like the larger one pictured here, to grow one or more extensions pointing to the entrance. The flag on the smaller helictite points in the direction of winter winds. We can assume that this stalactite receives its feed water mainly in winter, which is unusual.

89/STALACTITE

This stalactite, approximately 150 cm/59 inches long, is exceptionally large for an alpine cave. It has probably been growing for more than ten thousand years. It is growing at Holes-in-the-Floor passage, beside a narrow sedimentary dike that appears as a fringe of protruding rock to the left. The feed water is probably entering at the weak contact between limestone and dike rock.

90/CALCITE COLUMN

This column is the most massive single deposit of growing calcite yet discovered in the cave. It is nearly 3 m/10 feet high and more than 1 m/39 inches around. Its feed water is descending via the limestone-dike contact at the southeast end of Holes-in-the-Floor.

The small discoloured stump of calcite preserved on the ceiling just in front of the growing column is not as impressive, but it tells us more about the history of the caves. This is the remains of a large stalactite or column similar to the one we see today. It was eroded off nearly flush with the ceiling during one or more periods of protracted flooding. Clay carried by the floodwaters has impregnated the remaining calcite and turned it brown. There are many similar old, highly eroded deposits in the Central Grottoes, which indicates that the cave has suffered many cycles of erosion, deposition and quiescence.

91/FLOWSTONE BANK

The calcite encrustations covering a bank of clay in this scene are known as "flowstone." The bulbous white form above them is a "boss." There is a small stalagmite at the base of the flowstone.

Most of the deposits in the Central Grottoes are a dazzling white. This is quite rare in the world's caves. Pure calcite is white but most stalactites, stalagmites and flowstones are usually off-white, with distinctive tones of yellow, brown or red, due to minute amounts of organic acids within the calcite crystals. These acids usually come from decayed plant matter as the feed water passes through humus and soil with plant roots.

No plant material has been present above the centre of Castleguard Cave for tens of thousands of years, so the calcites are pure white.

92/CAVE PEARLS

Cave pearls, like pearls in oyster shells, are formed around a particle of sand or grit. In a cave, water supersaturated with dissolved limestone may drip regularly into shallow pools that contain grains of sand. The dripping action agitates and rolls the grains and precipitates calcite onto the rolling fragments. This builds up a succession of spherical sheaths, rather like building a ball of clay by rolling it over a clay surface.

Spherical and semi-spherical cave pearls are quite common in well-decorated caves. Usually they occur in clusters or "nests." A pearl rolled by a falling drop of water will displace its neighbours, causing them to roll as well. This interaction keeps the pearls in a nest growing at similar rates.

The Central Grottoes contain thousands, probably tens of thousands, of cave pearls in hundreds of nests. Most of the pearls are no larger than pinheads, but some, like those shown here, are 1 to 2 cm/½ to ¾ inch in diameter.

Pearls will stop growing if they are knocked out of the nest by a water drop or over-agitated neighbour, or if they grow too big to be rolled. Large pearls become cemented to the pool floor, parts of the growing flowstone. A new generation of tiny pearls may grow on top of them. Pinheads, displaced pearls and overgrown pearls are all shown in the photograph.

93/CUBICAL CAVE PEARLS

Nests of spherical cave pearls are quite common, but this nest of regular, cubical pearls is rare indeed. We know of only three other instances: in France, Puerto Rico and Romania.

This nest occurs on a ledge halfway up the Second Fissure. It is about 15 cm/6 inches in diameter and contains approximately three hundred pearls packed regularly in two layers, like sugar cubes in a box. Within 50 cm/20 inches of this nest are four other nests containing the normal spherical pearls. Some cube pearls have been thrown to the side of this pool and we can see others overgrown in its calcite floor. Four round pearls have been ejected into the cubical nest.

The cubical pearls are 5 to 7 mm/$\frac{1}{5}$ to $\frac{1}{3}$ inch in diameter. When a polished slice of one was viewed under a laboratory microscope it was seen that growth began with normal spherical shells of calcite. At a diameter of 1.3 mm/$\frac{1}{16}$ inch, it became crowded by its neighbours, and its sides grew progressively flatter where it was touched. We must suppose that the feed water dripping into this particular pool did so with precisely the amount required to shake the pearls but not enough to roll them once they had exceeded about 1 mm/$\frac{1}{20}$ inch. This is a remarkable example of the great stability that can exist in the underground realm, for the water drops must have exerted precisely the same force on the pearls for many years.

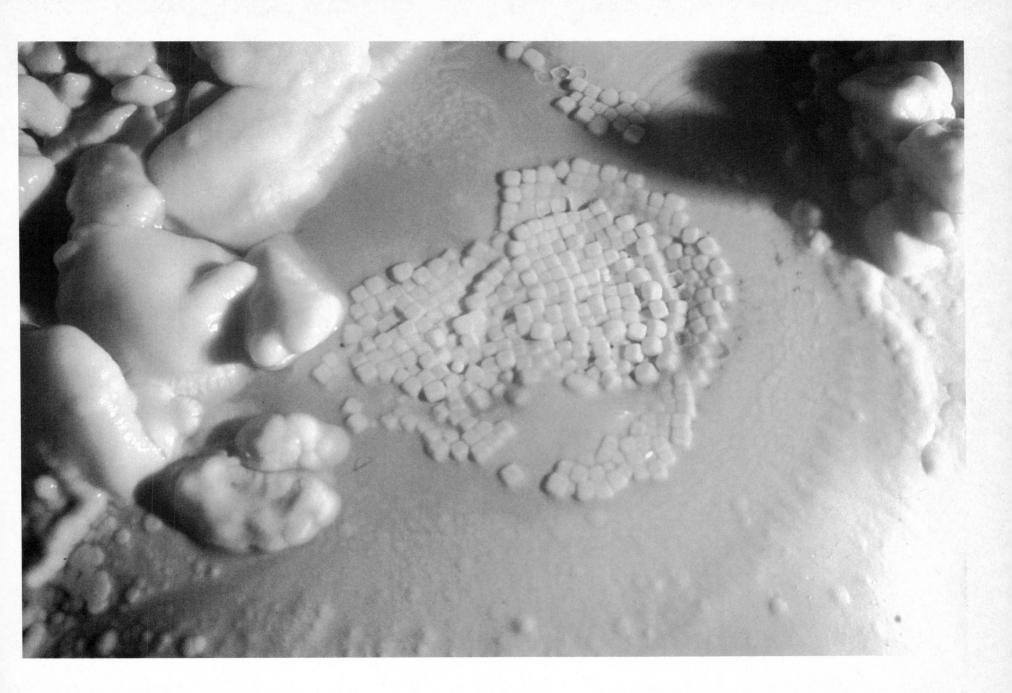

These two pictures have been juxtaposed to show how similar phenomena can occur in very dissimilar situations. Both are rounded, shield-like precipitates of carbonate that have been extruded from bedrock, and each is a few centimetres in diameter. There the similarity ends.

This bottom photograph is of a portion of the cave roof in the Central Grottoes. It lies almost directly beneath the site of the top photograph, but there are 300 m/980 feet of rock between them. The white deposits are a rare combination. The central, bulbous material is the lustrous mineral aragonite, which contains the same chemicals as calcite but has a different crystal structure (its atoms have a different geometric arrangement). The aragonite, rare in cold caves, is unstable; after a few hundred or thousand years the crystal structure changes to calcite.

Surrounding the aragonite is a cauliflower-like powdery deposit of hydromagnesite, a mixture of magnesium, carbonate and water $(Mg_4(CO_3)_3(OH_2).3H_2O)$. We are not sure exactly what causes this deposit, but probably water holding dissolved limestone and dolomite is extruded from the rock under pressure. The abundance of magnesium in the dolomite causes the calcium carbonate to deposit most of its film of water as aragonite rather than as calcite. At the aragonite perimeter the water is slowly evaporated by the cold draught that blows through the Central Grottoes. The exotic crust is created when some of the evaporating water is trapped by magnesium and carbonate molecules.

In the central cave the dark roof is decorated with many thousands of these mineral flowers.

Above the cave, on the Pika limestones on the south benches of Castleguard Mountain, are remarkable sub-glacial precipitates. Here, meltwater under pressure was forced down into the bedrock, where it flowed horizontally for some distance before returning to the surface through tiny holes. In the top photograph we can see that at each opening a hollow stalagmite has been built up, with an apron of calcite spreading outwards from it. These small deposits, 2 to 5 cm/1 to 2 inches in diameter, receive calcium carbonate from freezing water. They are a cold-water analogy to the large carbonate mounds and terraces that build up around hot springs such as Rabbitkettle in Nahanni National Park Reserve, Northwest Territories, or the Old Faithful geyser at Yellowstone Park in the United States.

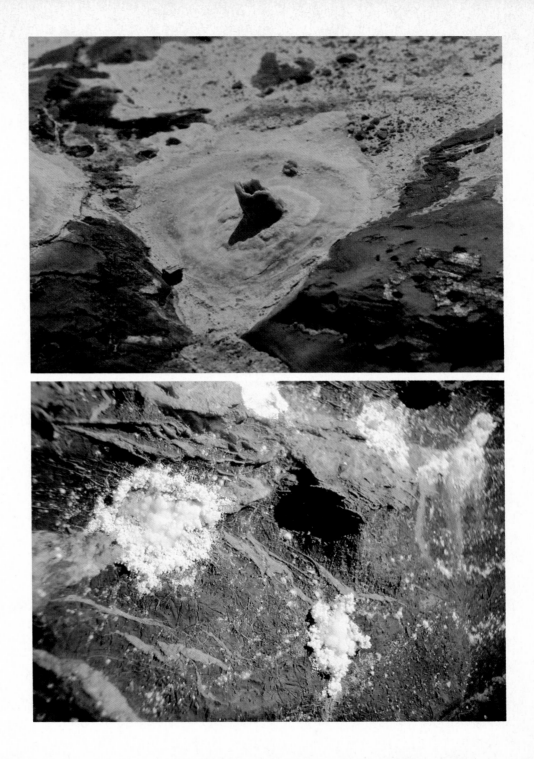

95/MICRO-GEYSERS

Surface micro-geysers form in series and groups, associated with pressurized channels in the rock by which meltwater arrived at the rock surface. All of the micro-geysers have been found on recently exposed benches shown in photographs 40 and 41, on the south side of Castleguard Mountain.

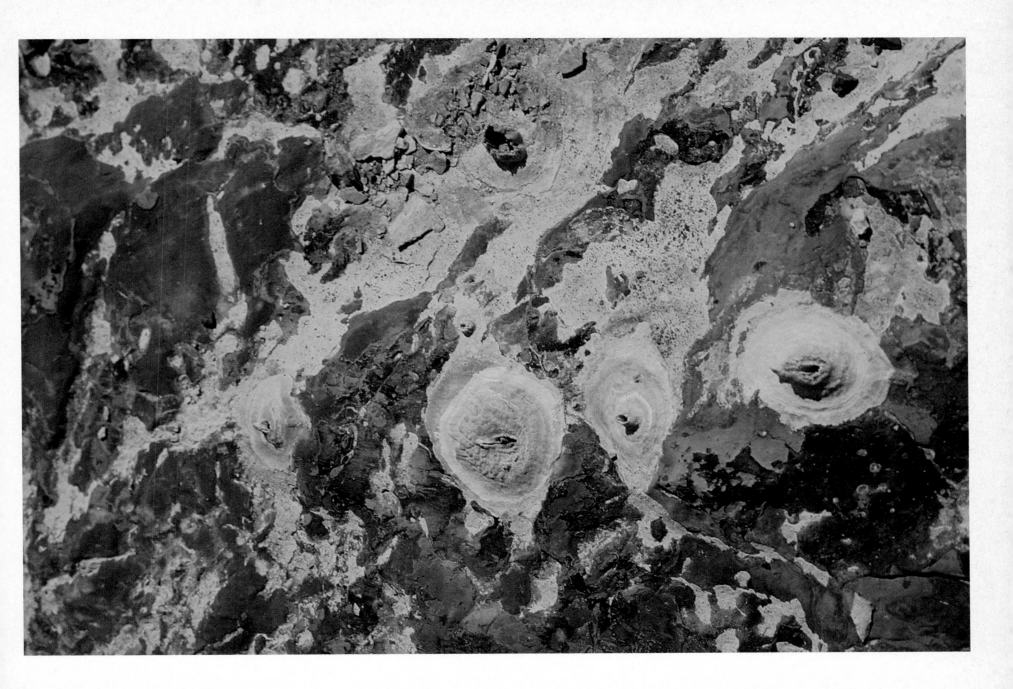

Castleguard Cave contains many deposits of silt and clay. (Evidence that they were quite deep in the past can be seen in photograph 75.) They are easily eroded by showering and trickling streams, and in most cases they are reduced to patches along the walls or to fragile bridges as in Holes-in-the-Floor.

The clay surfaces are crumbled, discoloured by dust and dull; yet when a face is cleaned, it reveals a rich, rhythmic pattern known as "varves," which can be read like the rings on a tree.

Varved clays are well-known in lakes fed directly by glaciers. In summer the lake is free of ice and fed by turbid meltwater. It is kept moving by melt streams and wind, so that only the biggest, dark-coloured particles settle out. In the winter, when the lake is still beneath an ice cover, the finer, lighter particles slowly settle to the bottom. Each varve, therefore, is a couplet of dark summer material overlain by lighter winter mud, and represents one year's accumulation.

In the varved deposits shown here the thickness of the couplets varies, suggesting that waters were more turbid at some times than at others. The prominent dark layer at the centre suggests some accidental event, perhaps a collapse of clay higher up the passage.

Varved clays are found throughout the cave, except in the modern flood zone. They appear to have been deposited during at least three different periods. Between these periods, the clays were eroded and stalactites and flowstones were formed. One varve is rarely greater than 1 cm/⅖ inch thick. During one period, varves accumulated in the central cave to a depth of more than 5 m/16 feet, representing perhaps five hundred years of continuous deposition.

During the principal ice ages, all of the Castleguard region except the highest mountain ridges and summits was buried by flowing ice. Springs draining the cave were blocked or back-flooded by water in the glaciers, and the cave became a deep lake under the ice. The clays settled from it.

This stalactite looks like a tree trunk that has been damaged and partly healed; in fact, its history is not dissimilar. It is an example of re-solution, a common feature in many caves, especially those that undergo severe climatic changes.

Supersaturated feed water deposited calcite to form a carrot-shaped stalactite. Then the chemical balance of the water changed for a time, becoming slightly acidic. It dissolved a deep cut the full length of the stalactite. The water is no longer acidic, but is once again supersaturated; a new, whiter calcite deposit is filling the top of the cut.

This is the ruins of a great flowstone boss in the Waterfall Room of the First Fissure. (Adjoining deposits are shown in photograph 79.) The boss was created by alternating processes. First, calcites were deposited on the rock and then were partially eroded by acidic water. More calcites were deposited, and some were eroded away. In each cycle, less calcite was eroded than deposited, until a large flowstone resulted.

Later, a great shower of acidic water from an overhead shaft developed. This water has dissolved the centre of the mass and the flowstone remnants are discoloured and pitted by splashes and runnels. Nevertheless, they are still more than 4 m/13 feet high and exceed 1 m/3 feet in thickness in places.

We are able to determine the age of many stalagmites and flowstones by uranium dating or by measuring their magnetism. This is one of the oldest deposits in Castleguard Cave. The oldest surviving parts are between 730,000 and 1,250,000 years old. The action of the acidic water shower began less than 50,000 years ago.

This calcite boss could not have begun to grow until the First Fissure in which it stands had been eroded to at least three-quarters of its present depth. The cave, therefore, must have been nearly fully developed 750,000 years ago.

99/GLACIER-ICE BLOCKAGE

Our traverse of the length of the principal passage in Castleguard Cave has been completed. In the entrance flood zone we were confronted with seasonal ice; here, permanent glacier ice has been injected into the cave, 10 km/6.2 miles distant from and 380 m/1200 feet higher than the entrance.

The farthest accessible complex of passages was first entered by a climb out of the Second Fissure in 1973. At this point, 800 m/2600 feet beyond the climb, the explorers were halted. The upper photograph was taken in April 1974, the lower one in April 1983: the ice front has not changed by more than 1 to 2 cm/1 inch during the nine years. The ice completely blocks the passage: it is a water- and airtight seal. It is supporting the block with the surveyors' mark, P1, upon it; in the past it pushed the blocks in the foreground to their present position.

The ice is opaque and coarsely crystalline. It has the same composition as glacier-ice samples taken high up on the Columbia Icefield, and there is little doubt that this is ice from the base of the icefield and has been extruded into the passage. Glacier ice is plastic and may flow slowly into an open space, as has happened here.

It is unlikely that there is more than 30 to 40 m/ 97 to 130 feet of cave passage beyond the ice front. Above the cavers' heads is perhaps 10 m/33 feet of rock under 230 m/760 feet of flowing ice of the main Columbia Icefield.

No other natural passage was known to be blocked by glaciers until the 1983 and 1984 expeditions to Castleguard Cave discovered six smaller passages terminating in ice. The cave has had its head cut off by a glacier.

These photographs emphasize the great stability of environmental conditions in the caves. No surface-glacier ice-front anywhere, even in the coldest part of Antarctica, would change as little as this in nine years.

100/TRAVERSE

It was feared that the glacier ice that sealed the passage beyond the Second Fissure would put an end to prospects for big new discoveries at the head of the cave. Yet, off to one side of that passage was a pit. This was descended in 1974 and found to end in rubble 23 m/75 feet below. In 1979, cavers crossed the Traverse, the top of the pit. A bolt was placed in the wall above the centre of the pit; a short ladder was hung from it. The caver clips onto the safety line and steps from the near edge to the ladder to the far edge.

This traverse opened up several more kilometres of passages and some deep shafts, all beneath the icefield. In 1983, five bolts were placed in the rock so cavers could cross another yawning pit, 60 m/200 feet deep.

101/BOULEVARD DU QUÉBEC

The Castleguard Cave exploring teams have always had members from across Canada and other countries. Groups of two or three cavers from the team take turns exploring promising new leads. First to cross the Traverse (photograph 100) was a group from the Société Québecoise de Spéléologie. From it they climbed into this beautiful gallery, which they named the "Boulevard du Québec."

It is a fine example of a composite gallery, the top an elliptical phreatic passage, a broad trench carved below. The trench walls have been shattered, perhaps by the load of glacier ice overhead.

There are many calcite decorations in this part of the cave. Most are much smaller than in the Central Grottoes, but some calcite crusts are visible on the roof.

102/STILLICIDE

Beyond the Boulevard du Québec (photograph 101) the passages divide and shrink until they become too low for cavers to explore.

This is an excellent example of a phreatic passage. Although it is probably very old and is only a few metres below the great glacier, the roof is a smooth and undamaged solutional form. Silt laid on the floor during the last flood is now dry and dessicated. Rubble from a side passage appears to have spread across it, perhaps sliding periodically when some meltwater seeps in.

The end of the cave is a very remote and utterly quiet place. It will take the caver two days to return to natural light. Thomas Hardy imagined "a lone cave's stillicide." This is it.

Return to Light

The great cave is part of the natural order of Castleguard. Here, much of the water system travels underground through a landscape that has never known light. When water from the cave rises to the surface, it joins water that has drained above ground. These blended waters eventually run into the ocean, where they deposit the rock and silt they have carried during their journey. Pure water – snow and rain – continues the process.

103/A FLOOD BURST FROM THE CAVE

These pictures were taken within a period of thirty seconds in the floodwater channel 60 m/200 feet downstream of the entrance to Castleguard Cave. The entrance is in the small cliff, partly concealed by trees at the rear of the scene.

The first 1.5 km/1 mile of the cave floods frequently in the summer. Two or three consecutive days of fine weather evidently create enough meltwater to fill the inaccessible Castleguard II and III systems underlying the downstream end of the known cave. Water then rises from flooded shafts and spills into the passages that lie below the P8

shaft 100 m/330 feet inside the cave. Some of this water drains to lower springs but the water level rises steadily up the shaft.

The peak time of glacier melting is about 2:00 PM. The floodwaters travel underground and usually rise in the shaft between 4:00 PM and 7:00 PM. Once over the lip of the shaft, they burst from the cave in a dramatic wave.

During July and August, under the right weather conditions, a daily flood peaks late at night and dries up soon after sunrise. In very hot weather there may be continuous flooding for many days.

104/THE BIG SPRINGS

At the peak of summer melt there are more than a hundred flowing springs in the Castleguard River valley below the mouth of the cave. All dry up in the winter except those few that are lowest in altitude. The springs are draining the heart of the Columbia Icefield, the upper Saskatchewan Glacier, Castleguard Mountain and Castleguard Meadows, via Castleguard II and III cave systems. Together they discharge 20 m³/710 cubic feet per second in fine weather.

The Big Springs are a group of three springs spaced about 100 m/330 feet apart that emerge in the north flank of the valley, 40 m/130 feet above its floor and 270 m/890 feet below the cave. The largest of the three is shown here at full flood. The waters are bursting from a narrow slot under considerable pressure.

In many years this spring flows continuously, from late June to mid-September. The water temperature is always between 1.5 and 2.2°C/35 to 36°F.

105/THE RED SPRING

This is how a cave begins. Water is pouring from a wide horizontal slot (a bedding plane), but the opening that has been eroded is no more than 2 to 3 cm/1 to 1½ inches high. As hundreds and thousands of years go by the channel will deepen and the slot will become a cave passage several metres wide and deep.

The spring has been named "Red Spring," from the distinctive colours of the mosses and algae that grow on the cliff below it. Most springs in this region dry up during the winter, but this one flows throughout the year, dwindling to a litre or two per second, less than one percent of the July flow seen here. The water is coming through a floodwater channel from Castleguard Cave, 150 m/500 feet to the rear of the picture, and it is believed to drain parts of the southern half of Castleguard Meadows.

106/THE BIG SPRINGS' WATER REACHES THE VALLEY FLOOR

The Big Springs are seen at the bottom left, where they begin a 40 m/130 foot plunge to the valley floor. Young springs are bursting from slits in the rock face beyond it. Their water is racing away in the nearer channel. The distant stream is the Castleguard River, which flows from the snout of the south glacier of the Columbia Icefield. In the gravel flats between the two channels there are some twenty more small springs, which are believed to come from the same source as the Big Springs.

107/WATERSMEET

This is the end of the karst ground-water system at Castleguard. Water from the Big Springs on the left merges with Castleguard River on the right, approximately 300 m/984 feet downstream of the Big Springs (photograph 104).

Castleguard River is a glacier melt stream that flows on the surface from the snout of the south glacier, Columbia Icefield. It carries large amounts of "glacier flour," suspended matter that gives it a milky appearance. The Big Springs' water also originates in the Columbia Icefield but comes through the caves under Castleguard Mountain. It carries little glacial flour and is nearly clear.

It is evident that the Big Springs are bigger than the river, which is being pushed to the right bank when the waters meet; but within a few hundred metres all the waters will be mixed and milky.

From the sinks on the Columbia Icefield and the ice blockages in Castleguard Cave beneath, to Watersmeet in the valley far below, the cave has rewarded a generation of cavers. Between these limits there is more cave, maybe much more. Little may ever be explored, because most of it is likely too confined for the human body or is filled with very cold water. But youth presses on, and potential areas for discovery are constantly being studied.

The Experience

An unfamiliar landscape broadens perceptions. Castleguard is stark simplicity and rich complexity, fragility and stability, subtlety and overpowering presences.

108/SEARCHING

Most societies espouse ritual contemplation in solitude, but the concept is more distant now and solitude less accessible. Nonetheless, we still have the need to withdraw sometimes, to get away from it all, however briefly, for private contemplation.

Perhaps the hard rocks and ancient mountains, the exhausting climb and raw scenery are reassurances of a reality here at Castleguard. Perhaps, too, the softer qualities of water and nearby life forms are comforting. And in most cultures there is a counterpart to this image of a free-searching spirit in the wilderness.

109/FINDING

We search open spaces for exaltation, nooks and crannies for hidden secrets. Behind an idyllic waterfall lies refreshing coolness; a thin veil briefly separates personal space from Castleguard beyond.

110/SERENDIPITY

Within enduring landscapes lie ephemeral things. First-hand experience of them is highly prized because they will never happen again in quite the same way.

Like the air bubbles in cave ice that form patterns that are never repeated, the explorer discovers the uniqueness of human experience.

The experience of Castleguard is a journey through time.

The rocks were formed hundreds of millions of years ago. The mountains are younger, pushed up by forces some tens of millions of years ago. The great glaciers passed over them repeatedly within the past few million years and just melted away about ten thousand years ago. The Little Ice Age happened within history, during the past seven hundred years. It ended during this century, within living memory. The meltback continues today. Each year winter passes over the land and changes the details of Castleguard. Late-spring storms delay the onset of growth and summer frost may end the floral display on the Meadows.

At Castleguard, we can read time in the landscape. Beneath all that is visible on the surface lies another landscape, the Cave; it too is a mirror of time.

Castleguard is a perfect embodiment of the National Parks concept: a heritage of important natural landscapes preserved for all time, while making possible recreation, discovery, insight and memories.

Early adventurers dreamed of streets paved with gold. We have built a civilization that surpasses this crudely imagined luxury. Still, we search, and sometimes find a greater treasure.

The sun that burned us through the day had risen behind Terrace Mountain and flooded over the glacier nearby. At sunset it turned the opposite valley wall to gold against a threatening sky. Then, it swept the shadow of another mountain over us, over the trees beyond the stream, the valley, the huge moraines and the high valleys. It lingered briefly on remnant snows and ice that still tear at the mountains before darkening the peak.

Most of us live so far from a glacier that it is nothing more than a remote object of curiosity. Yet all of Canada is a glaciated landscape. We obtain food and wood from its soils, live on it and bury each passing generation in it. As we stare past the warm glow of the fire at the sunset on the shrunken glacier across the valley, we can experience this continuum. The small, unnamed glacial remnant on Terrace Mountain, and our near one that is warmed by the rising sun, are realities that confirm our place in the natural order and link us to processes still moulding the face of earth.

An Overview

Most people who live in Canada forget that our landscapes developed from the glaciers and melting patterns of the great ice age. But less than twenty thousand years ago, most of the mountains in western Canada were covered by ice thousands of feet thick; the rest of the country was buried even more deeply. The Columbia Icefield, today an isolated and stark reminder of the ice age, was once indistinguishable from the glaciers that covered the continent. The last ice age ended about ten thousand years ago, leaving only a few small remnants in the Arctic and the highest western mountains. Among the largest of these is the Columbia Icefield, 325 km²/127 square miles of glacier ice. It lies across the Continental Divide, where three watersheds meet. Meltwater flows in three directions: into the Arctic Ocean, into the Pacific Ocean and into Hudson Bay.

The icefield is relatively stable, but small arms of it, called valley glaciers, move slowly down the surrounding valleys, where they eventually melt. The largest of these glaciers is the Saskatchewan, which flows eastwards from the central icefield. Next to the icefield, in Banff National Park, is Castleguard. The scenery at Castleguard has been changed since the last great ice age, but only in minor ways, by the brief Little Ice Age. By examining Castleguard, we can see the many processes that led to modern-day Canadian landscapes. Beside Castleguard Mountain, at the edge of the Columbia Icefield, a meadow and a wide, horseshoe-shaped valley are suspended high between two mountains, cut off at one end by the Saskatchewan Glacier and at the other by a river gorge. The valley is not named on maps. By tradition, the whole landscape is referred to as "Castleguard."

The surface of the valley was formed some ten or twelve thousand years ago, when the last or Wisconsin Glaciation ended. Land-forms and vegetation attest to the valley's age and origin. Other modern Canadian landscapes were also being formed at this time.

Beneath the Castleguard Meadows is a vast cave complex. The cave has only one entrance, in the steep slope that terminates at the lower end of the Meadows. The system runs north. Exploration has been pushed beneath the icefield, well beyond Castleguard Mountain, where progress is stopped by ice blocking the passages.

WAYS OF LOOKING AT THE EARTH

Geologists study rocks: their origin, age, chemical composition, physical qualities, sequence of formation, internal changes and the causes of these changes.

Geomorphologists take the rocks as a given and study land-forms, what happens to the rocks after they were formed. They examine the shapes of mountains, the distribution of valleys, the form of river channels, the scale of their meanders, the gradient of a beach and its growth or destruction over the years.

Vegetation may be viewed as the earth's clothing, and ecologists see it as time-related. It is a major component of landscapes, and brings about changes in the appearance of physical landscapes and the rate at which many physical processes operate.

PHYSICAL LANDSCAPE CHANGES

Geomorphologists recognize that, as the earth's surface is altered, different land-forms develop in different climates and upon different types of rock. Original constructional features, such as folds and faults in the rocks, may determine the size and shape of land-form units, but they differ now because of erosion. Four principal kinds of erosion work upon the continents.

The most widespread is the "fluvial" or river system, the most common erosion system acting in most parts of Canada today.

The second system is the "arid" (or warm-to-hot-desert) system. This kind of erosion is characterized by slow chemical rotting, sporadic sheet floods and wind. The arid system is not important in Canada, although the wind has moulded some of the arctic plains and small dune fields, deltas and beaches.

Around the margins of the land, a third system, wave attack and water currents, erodes platforms and cliffs, and creates beaches, bars and spits of sand and shingle. This system is not important at Castleguard.

The fourth principal system is caused by glaciers. Bedrock is eroded and loose material is moved by the flowing glacier ice. This system is a central theme of this book. Twenty thousand years ago, thirty percent of dry land on earth was covered by glacier ice. Today, only ten percent of the land is under ice. (Most of it is in Antarctica and Greenland.) Except for a few bits of the Yukon, all of Canada was buried. The sheet of ice that covered North America was larger than Antarctica. Canada's Rocky Mountains were buried, except for a few peaks and high ridges. The ice receded slowly – it took about seven thousand years – and was gone about seven thousand years ago. The age that started when the ice left is called the "post-glacial" or "Holocene" period. The size of the earth's glaciers has waxed and waned many times in the Pleistocene period, the geomorphologists' term for the past two million years or so.

To a geomorphologist, the landscapes of Canada are dominated by effects of glacier erosion during the many ice ages and by deposits of debris left by glaciers. In the mountains, glacial-erosion forms predominate. The flowing ice scoured valley floors and undercut rock ridges and peaks; it produced the sharp relief called "alpine topography," the most rugged kind of landscape on our planet.

Now that most of the ice is gone, flowing water is the main agent of erosion and debris transport in the Canadian Rockies. But in geomorphic terms, very little time has elapsed since the last ice age. The largest mountain rivers have only had time to retouch glacial landscapes or redistribute debris locally.

There are two kinds of glacier landscape at Castleguard. In the Castleguard Meadows we find beautifully represented "relict" glacial landscapes. But near the Castleguard Mountain we can see modern glacier landscapes created by the Little Ice Age.

THE LITTLE ICE AGE

About seven hundred years ago, there began a slight but significant cooling of the climate in many northern regions of the globe. Surviving glaciers began to thicken a little and to advance a few miles down the valleys in which they were resting. This expansion was trivial in comparison with a full ice age; it is known as the "Little Ice Age," or the neo-glacial period. The Little Ice Age began to end about a hundred years ago, when mountain glaciers began to thin and recede at the margins. The glaciers shrank quite quickly between about 1930 and 1960, and they are still shrinking. Clear evidence of their retreat can be seen around most glaciers in the Rocky Mountains, especially in the southeast quarter of the Columbia Icefield, where Castleguard Mountain stands.

In high valleys close to the mountain, we can see dead trees, their root systems damaged by the movement of ice during the Little Ice Age, leaning against the undamaged live trees of a mature alpine forest. We know from photographs that the dead trees were killed about eighty years ago.

As the glaciers retreat, successive waves of vegetation are still advancing onto the newly exposed land.

There are many active glacial and river landscapes in the western mountains. At Castleguard, two other distinctive erosion systems compete with them; the story becomes especially rich and complicated.

FREEZING PROCESSES

One of these systems is alternating frost and thaw. The process operates on bedrock, debris piles and soils. It is often called "the periglacial system" because it occurs near glaciers or is affected by glaciers. The periglacial system shatters rock into rubble. A crack in a rock is filled with water; the water freezes and expands; the rock shatters. Then the shattered rock is formed into patterns of polygons, stripes or lobes by the ice that lies in its interstices. As nearby glaciers melt, the soil and rubble flow slowly down slopes saturated and lubricated by meltwater. This slow flow is termed "solifluction." The rubble flows most smoothly down slopes that are permanently frozen (permafrost) below the surface. Only the top layer of the slope will thaw.

Periglacial land-forms are widespread in arctic Canada and in the subarctic mountains of Labrador, the Northwest Territories and the Yukon. Often, the delicate patterns carved by the periglacial system appear on top of earlier glacial land-forms. In the Rocky Mountains, the periglacial system operates mainly above the tree-line. In forested parts of the Meadows there are few periglacial effects; in the higher meadows we see numerous periglacial patterns that dominate on high ridges; as well, some slopes are characterized by spreads of frost rubble.

KARST PROCESSES

Solutional land-forms, called "karst," are the second distinctive landscape system represented at Castleguard. Karst land-forms are created when rock dissolves in water. Acidic water, from rain, snow or glacial melt, sinks underground and enlarges solution channels, which coalesce to form natural plumbing systems. The water emerges at springs lower down. The karst system differs from river (or "fluvial") systems in that cave plumbing is all underground. The karst system develops best in limestone and dolomite rock. These make up most of the landscape at Castleguard.

As time passes, more water carves its way through the rocks and the caves grow larger. Sometimes whole series of large passages in a system may be left drained and abandoned, or "relict," as the water finds new conduits beneath them. The explored cave at Castleguard is mostly abandoned passage. From the downstream end, where explorers enter, to its ends upstream, explorers have mapped 18 km/11 miles of passages. These passages rise 373 m/1225 feet from the entrance to the far end.

Beneath the cave passages that have been explored are at least two other great systems of cave passages. These two lower systems, called "Castleguard II" and "Castleguard III," act as drains for the summer run-off and supply water to the Big Springs at the southeast end of the karst. Despite some brave efforts, no one has entered Castleguard II or III; the passages are too small for people to enter, or they are filled with water. We know something about these systems, though, because we have studied the water entering them via sinkholes and have gauged their output at the Big Springs. Castleguard II and III may contain more than 100 km/60 miles of passages.

The explored cave extends far under the Columbia Icefield. Meltwater from the base of the glacier drains directly into the lower, unexplored cave. Some of the farthest passages of the explored cave are blocked by glacier ice, which forms airtight and watertight seals, and may have less than 10 m/33 feet of rock above them; on top of the rock is more than 230 m/750 feet of glacier ice. Other passages branch into smaller tubes, which supply air to the glacier bed in winter.

Karst land-forms also develop at the earth's surface. Run-off waters drain underground via two types of surface land-forms, karren and sinkholes.

Karren is a collective German word for a great variety of small solution forms on rocks. Larger karren are fractures and clefts that have been widened and deepened by solution. When many of them intersect in a surface, they are called "grikes." The blocks of rock between the grikes are known as "clints." Clint-and-grike topography may be so geometrical and regular that it looks man-made, and is often called "limestone pavement." This land-form is well-developed at Castleguard.

"Sinkholes" are large water-swallowing points, as the name implies. The most common types are funnels, or bowl-shaped depressions, that drain water to a central "soakaway," which is usually covered with soil. Sinkholes are sometimes 1000 m/3280 feet in diameter. Alpine terrains, such as Castleguard, have small, steep sinkholes with many vertical shafts.

Karst features are abundant at Castleguard. In the underground cave, individual passages maintain their form and dimensions for exceptional distances. In the middle of the cave, underneath the mountain, there are beautiful stalactite and stalagmite displays where there are good scientific reasons for *not* expecting them. On the surface are the world's finest displays of sub-glacial calcite precipitates, the product of combined karst and glacial processes.

Glacial, karst and frost processes compete. Glaciers scour karst forms; sinkholes drain glaciers to diminish their scouring power. Water infiltrates and prepares karst rock and frost shatters it. Many river-system and surface-karst features were destroyed during the last ice age. (A few survived in modified form and are still functioning.) Underground plumbing is more protected from glacial disruption than are surface forms. In the relict caves at Castleguard, some of the longest glacial and interglacial sequences in the Rocky Mountains are recorded.

Most natural landscapes develop very slowly. At Castleguard, we can see both recent and ancient examples.

ECOLOGICAL LANDSCAPE CHANGES

Ecology has been defined as the study of the "household of nature," how living things relate to their environment and to each other. (Biology is the study of life itself, origins, evolution, forms, basic processes and properties.) Ecologists are interested in single species and in communities that contain many species, which interact. They have also noted that life forms and communities tend to respond to change in a way that seems out of proportion to the stimulus for change.

Autecology concentrates on the individual and its relations to everything else. Synecology looks at populations, with their larger matrix of possible relations, both mutual and external. The distinction is not as great as these two formidable words might imply, and we use both points of view to describe the situation at Castleguard.

VEGETATION

Using energy from the sun, plants convert atmospheric carbon dioxide (a gas) and water (a liquid) into a biologically useful "food" called carbohydrate. This conversion system has a by-product: dead plant material enriches the soil and improves its physical properties, encouraging further plant growth. Decaying vegetation also acidifies the soil, and any water percolating through. At Castleguard, mats of dead plants, called "peat," affect drainage of the landscape on a small scale, and acid waters speed cave-forming processes in underlying limestone bedrock. (In other parts of the world, organic terrain is a dominant landscape type, and an important economic resource.)

Vegetation is the living world's response to landscape. The contest is a simple one: living things struggle to occupy all suitable habitat. We look more closely, therefore, at the types and patterns of vegetation than at individuals and species, and we measure their success (or lack of it) by the amount of available terrain they cover. Many examples,

ranging from complete failure to overwhelming success, can be found at Castleguard. Pretty flowers are a delight to the eye, but their ecological significance lies in their numbers.

In high mountains and in the Arctic beginnings of life are the real problem. Here, many ecologists concentrate on pioneer plant communities. More southerly regions do not have that problem, and studies there tend to concentrate on developmental processes and establishment of long-term stability in so-called climax communities.

VEGETATIVE COVER

Some parts of the Meadows at Castleguard have a complete vegetative ground cover of small alpine plants. These places usually have adequate soil and a reliable water supply, and they offer protection from the worst of the winter winds. A complete or "closed" ground cover usually means a prosperous community; dead plant material is retained on site, adding to soil richness, instead of being blown or washed away.

Like small perennials and herbaceous annuals, trees also struggle to establish and occupy a landscape, then struggle again to dominate, competing for sun, moisture and soil nutrients. Therefore, a "closed" forest is one in which there is no room for more trees. Seedlings may spring up but they must wait until the giants decline and pass from the scene.

Trees have another ecological problem, one not shared by small ground-hugging plants. In regions with cold winters, where the ground freezes and the snow is deep, root systems freeze and become immobilized. Long after spring warmth stimulates growth above ground, some tree roots are still frozen solid. The growing tops are exposed to warm weather, especially at high altitudes where the hot spring sun is reflected from the snow. The tree tops need water for growth, but frozen roots cannot supply it. Some trees die. This problem is called "physiological drought" and affects plants that grow in ground that freezes in winter. Trees are especially affected because they project far above the snow. In "temperate" regions this problem sometimes affects our lawns and some cereal grain crops; we call it "winter kill." Low temperatures alone seldom cause winter kill; it is the sudden onset of warm growing weather that does the damage.

NICHE

Directly or indirectly, every living thing interacts with every other living thing. First, an organisms need food and shelter for survival. Only after these needs are met can they interact with another life forms. Early ecologists coined the term "niche" to describe the many indefinable but important relationships that living things have with their surroundings. A niche is a place to live and a way of life. The term is imprecise but useful when making generalizations. Just as people and animals usually find a role in life, so a plant will prosper where conditions are suitable and disappear when its needs are not met. The need for a niche causes the sequence of plant communities, called succession, that occurs where ecological conditions are changing.

Two kinds of plants can establish themselves in newly available habitats. First are the hardiest primitive plants, such as lichens, which are true pioneering plants and need only a stable, non-toxic surface on which to develop. The second group may be more highly evolved but must be able to prosper in poor soil, without shelter or other favourable conditions. At Castleguard, both groups are present.

SUCCESS, CLIMAX, CHANGE

After the initial colonization, plant-community types tend to succeed each other in a reasonably orderly manner and evolve into a "climax" or self-perpetuating community that is stable over long periods of time. But if a plant community persists, its presence may change local conditions. If these changes do not suit the species, another community will develop. The definition and identification of climax communities is still debated among ecologists.

The climax concept assumes that environmental conditions do not change. Yet we know that climatic conditions do change. For example, the Little Ice Age must have affected the speed or direction of some plant sequences, and perhaps modified achievement of a climax. At Castleguard we have seen that the Little Ice Age killed some well-developed alpine forests.

In much of North America, ecologists say, there has been an orderly succession of plant communities since the last ice age. There is also evidence that successional stages and climax communities are moving north slowly. One good indicator is bird life. Birds are mobile, and they can occupy new habitat or abandon unsuitable habitat quickly. Bird enthusiasts report that many species are extending their ranges to the north, into formerly marginal environments. Few bird species have "withdrawn" southwards since bird-watchers began to keep records. The classic climax plant community may be more like a slow-moving wave that advances and stagnates as climate improves or deteriorates.

The Little Ice Age was a time of "difficult"

weather for most northern civilizations. At Castleguard, glaciers advanced down valleys where none had existed for five or ten thousand years. The weather must have set back the "ecological-succession" clock, but by how much we can only surmise.

Castleguard displays a complex pattern of Little Ice Age effects (a few hundred years old at the most) superimposed on last ice age effects that are perhaps ten thousand years old. We know how far the ice advanced, but we don't know how the local climate and weather changed, or what effect the changes had on vegetation and successional development.

Had we witnessed the events of the past seven hundred years, we would understand more clearly the implications of current environmental changes. We think that if the mean annual temperature changes by a few degrees – or possibly by only a single degree – there will be profound changes in plant growth. An environment such as Castleguard is a good place to study the effects of small temperature changes, because local conditions are very delicately balanced. A very slight cooling or warming can cause glaciers to advance down the valleys or to retreat so quickly that they leave only bedrock behind.

Castleguard has a fascinating and dynamic ecology. A more complete understanding of conditions there might shed light on the climatic changes of recent centuries and might help to interpret observed changes in the world today.

Plant communities, especially pioneering ones, develop slowly; sometimes it takes decades or centuries for only a few changes to occur. But plants respond much more quickly than do most physical landscapes where, typically, millenia are needed to realize significant change. This is the value of Castleguard: the landscape is changed almost in step with the vegetation. Castleguard now is a very different place than it was even a hundred years ago, when Banff, Canada's first national park, was established.

A SYNTHESIS

We have seen how physical processes have changed Castleguard. Great and small glaciations scoured the landscape. Acidic rain, meltwater and frost are reworking the surface. Waters disappear underground to work even greater change in total darkness.

We have observed, too, how vegetation established and developed, greened much of the valley and initiated growth processes that, during millenia, can bury whole landscapes deeply under dead plant debris. The debris acidifies local water and enhances the physical process that dissolves solid rock.

Thus the softest things – water, the stuff of winds and fragile life that arose from them – work separately and together, tearing apart mountains and hollowing them out. In the ponderous presence of glaciers and the elements of climate, we sometimes forget that life too is a landscape process.

At Castleguard we can read the debris of past events. We observe landscape processes as they happen: backdrop of running water, boom of falling ice, clatter of shattered rock. Where plants grow we can watch elemental survival.

In celebration of a landscape, we walk small upon its surface.

Topographic Map of the Columbia Icefield

The map shows the Columbia Icefield and the principal glaciers, mountains, rivers and passes of the region. The Continental Divide (Alberta-British Columbia boundary) passes through the centre; the Icefield Parkway (Banff-Jasper highway) crosses the northeast corner. The approximate course of Castleguard cave is indicated south of Castleguard Mountain. The inset locates the icefield in Alberta and British Columbia.

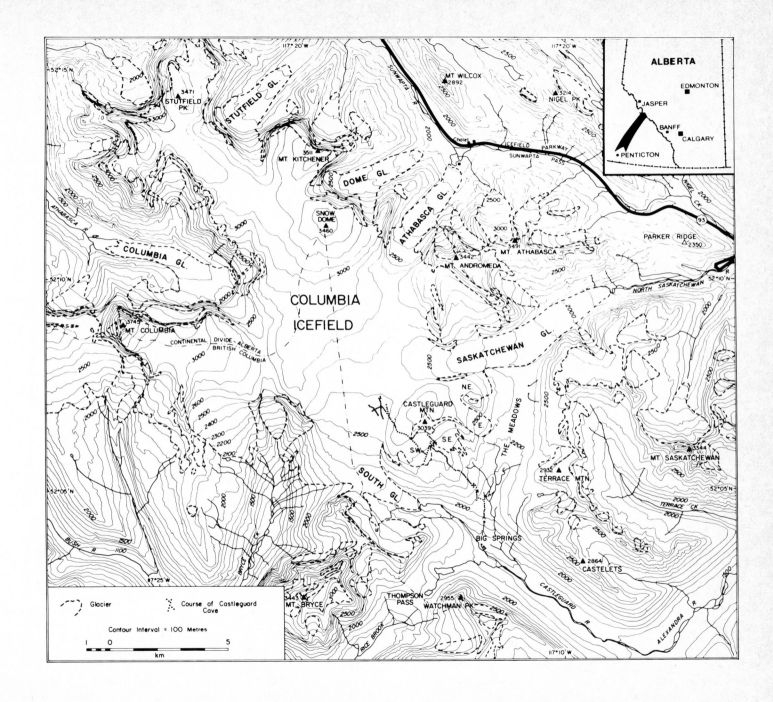

52°15'N

117°20'W

3471
STUTFIELD
PK

STUTFIELD GL.

MT WILCOX
2892

NIGEL PK
3214

3511
MT KITCHENER

DOME GL.

Chalet

ICEFIELD PARKWAY

SUNWAPTA PASS

93

SNOW
DOME
3460

ATHABASCA GL.

PARKER RIDGE
2350

52°10'N

COLUMBIA GL.

52°10'N

3000

3442
MT ANDROMEDA

3491
MT ATHABASCA

2500

NORTH SASKATCHEWAN R.

3749
MT COLUMBIA

COLUMBIA
ICEFIELD

CONTINENTAL DIVIDE — ALBERTA
BRITISH COLUMBIA

SASKATCHEWAN GL.

N.E.

CASTLEGUARD
MTN.
3039

E.

S.E.

S.W.

THE MEADOWS

2500

3344
MT SASKATCHEWAN

52°05'N

2935
TERRACE MTN

SOUTH GL.

BIG SPRINGS

TERRACE CK

2864
CASTELETS

52°05'N

BUSH R.

1100

1500

BRYCE R.

3445
MT BRYCE

THOMPSON
PASS

2955
WATCHMAN PK

CASTLEGUARD R.

ALEXANDRA R.

117°25'W

RICE BROOK

117°10'W

Glacier

Course of Castleguard
Cave

Contour Interval = 100 Metres

1   0   5
km

Castleguard and the Icefield

This map shows the distribution of the principal glacial and karst land-forms, ice, water and vegetation that appear in the photographs. The scale of important features, such as moraine ridges, has been exaggerated for emphasis. Only the general distribution pattern of karst sinkholes is suggested.

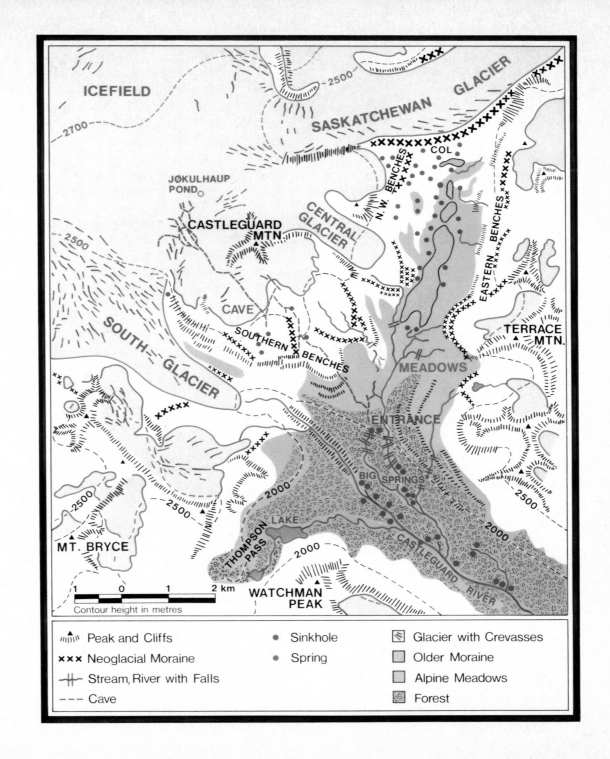

ICEFIELD

2700

2500

SASKATCHEWAN GLACIER

2500

COL

JØKULHAUP POND

CASTLEGUARD MTN.

CENTRAL GLACIER

N. W. BENCHES

EASTERN BENCHES

CAVE

TERRACE MTN.

SOUTHERN BENCHES

SOUTH GLACIER

MEADOWS

ENTRANCE

BIG SPRINGS

2500

2000

2500

2000

THOMPSON PASS

LAKE

CASTLEGUARD RIVER

MT. BRYCE

2000

2000

WATCHMAN PEAK

1   0   1   2 km

Contour height in metres

	Peak and Cliffs	●	Sinkhole		Glacier with Crevasses
xxx	Neoglacial Moraine	●	Spring		Older Moraine
⫞	Stream, River with Falls				Alpine Meadows
---	Cave				Forest

Cross-section of the Castleguard Karst

This is a schematic representation of a section extending northwest to southeast through the Castleguard karst. The vertical scale is exaggerated four and a half times to emphasize mountainous landscape. A simplified profile of Castleguard cave is shown in red. Watercourses shown are hypothetical, but the springs are real. "Waterfowl," "Arctomys," "Pika," "Eldon," "Stephen" and "Cathedral" are the names given by geologists to the principal rock formations.

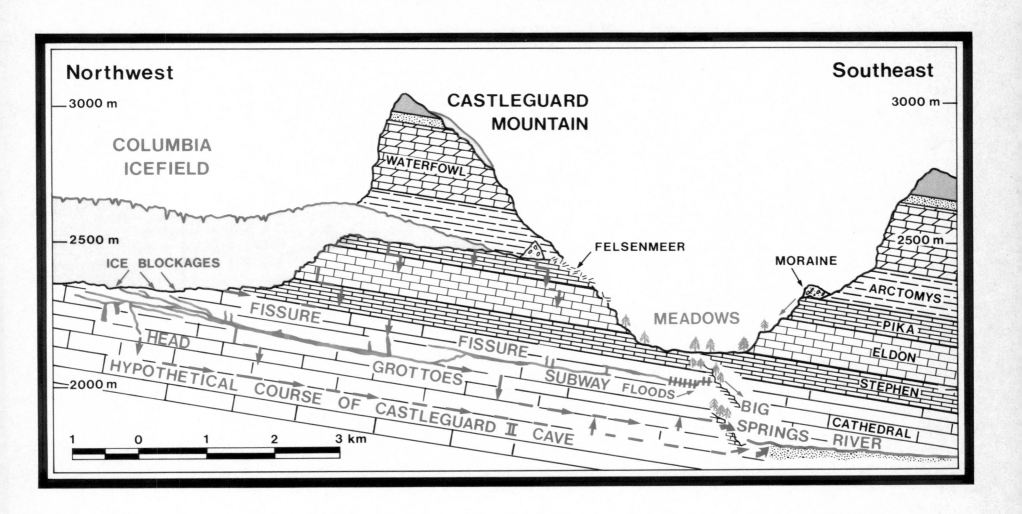

Northwest — Southeast

3000 m — 3000 m
2500 m — 2500 m
2000 m — 2000 m

COLUMBIA
ICEFIELD

CASTLEGUARD
MOUNTAIN

WATERFOWL

FELSENMEER

MORAINE

ARCTOMYS

PIKA

ELDON

STEPHEN

CATHEDRAL

ICE BLOCKAGES

FISSURE

MEADOWS

HEAD

FISSURE

HYPOTHETICAL COURSE OF CASTLEGUARD II CAVE

GROTTOES

SUBWAY FLOODS

BIG
SPRINGS — RIVER

1   0   1   2   3 km

Photographers

Daniel Caron	Photographs 93, 101
Derek Ford	Photographs 5, 34, 42-44, 46, 47, 49, 53, 57, 59-61, 65, 67-69, 71, 73-87 89-92, 94(bottom), 95-98, 103, 104, 106-108, 110, jacket (bottom)
Russell Harmon	Photograph 75
Stein-Erik Lauritsen	Photograph 99(bottom)
Thomas Miller	Photographs 88, 100, 102
John Mort	Photograph 70
Dalton Muir	Photographs 1-4, 6, 7, 9-28, 30-33, 35-41, 45, 48, 50-52, 54-56, 58, 62-64, 66, 72, 94(top), 105, 109, 111-113, jacket (top)
Anthony Waltham	Photograph 99(top)
Stephen Worthington	Photographs 8, 29

Maps drawn by the Department of Geography, McMaster University

The print of the photograph by A.O. Wheeler, inset in photograph 33, is held by the Geodesic Survey of Canada, in PARC BOX 1024.

Project Manager:	Susan Cargill-Johnson
Editor:	Charis Wahl
Designer:	Karen Okada

Acknowledgements

Derek C. Ford owes the deepest debt to his companions in the field at Castleguard. Since investigations began in 1967, more than seventy persons have worked very hard and safely in this place that is difficult, cold and dangerous, as well as beautiful. He especially thanks Bob Bignell, Michael Boon, George Brook, Charlie Brown, Julian Coward, John Drake, Ralph Ewers, Peter Fuller, Michael Goodchild, Russel Harmon, Bryan Luckman, Tom Wigley and the late Gary Pilkington.

Our understanding of the behaviour of underground water at Castleguard has been greatly enhanced by Chris Smart's hydrological research. The cave maps are from a 1984 revision of the survey directed by Steve Worthington. Many officers of Banff National Park, in particular Tommy Ross, have offered willing aid and good company over the years. Financial aid for research has come primarily from the Natural Sciences and Engineering Research Council of Canada (formerly National Research Council). Parks Canada has also been generous. Finally, the Science and Engineering Research Board of McMaster University, Hamilton, Ontario, has generously provided supplementary support.

R. Dalton Muir is grateful for the support of Canadian Wildlife Service in co-operation with Parks Canada, both agencies of Department of the Environment, Ottawa, for the opportunity to initiate and pursue this publication as part of celebrations marking the Centennial of National Parks in Canada.